Luay A.A. AL-Swidi
Auda S.A. Saeed

Soft Fuzzy and fuzzy soft set theory With applications

Luay A.A. AL-Swidi
Auda S.A. Saeed

Soft Fuzzy and fuzzy soft set theory With applications

Fuzzy Reliability

Noor Publishing

Imprint
Any brand names and product names mentioned in this book are subject to trademark, brand or patent protection and are trademarks or registered trademarks of their respective holders. The use of brand names, product names, common names, trade names, product descriptions etc. even without a particular marking in this work is in no way to be construed to mean that such names may be regarded as unrestricted in respect of trademark and brand protection legislation and could thus be used by anyone.

Cover image: www.ingimage.com

Publisher:
Noor Publishing
is a trademark of
International Book Market Service Ltd., member of OmniScriptum Publishing Group
17 Meldrum Street, Beau Bassin 71504, Mauritius

Printed at: see last page
ISBN: 978-620-0-77600-6

Soft Fuzzy and fuzzy soft set theory With applications

By
Prof. Dr. Luay Abd AL-Haine AL-Swidi
Prof. Dr. Audia Sabri Abd Al Razzak Saeed

Department of Mathematics / College of Education for Pure Sciences/ University of Babylon

2020

Table of Contents

List of symbols

Symbol	Description
$\tilde{A}$	The fuzzy set
$\leq$	The fuzzy subset

$\vee$	The fuzzy union
$\wedge$	The fuzzy intersection
$\in$	The fuzzy belonging
$\tilde{A}^{\circ}$	The fuzzy interior
$\bar{\bar{\tilde{A}}}$	The fuzzy closure
X	The Universal set
$P(X)$	The Power set of X
E	The set of parameters of X
I_i^X	The families of fuzzy sets where i= 1,2,3,…,7
$SS_i(X)$	The families of soft sets where $i = 1,2,3,4$
F_e^x F_x F_e	Type I soft point Type II soft point Type III soft point
F_A	Soft set
$F_A^{\ddot{c}}$	The complement of the soft set F_A
f^{τ}	The fuzzy topology
s^{τ}	The soft topology
$\mathrm{d}(F_A)$	The derived of the soft set F_A
$cl.(F_A)$	The closure of the soft set F_A
$int.(F_A)$	The interior of the soft set F_A
ext.(F_A)	The extensor of the soft set F_A

f.p.	Fuzzy point
s.p.	Soft point
f.s.p.	Fuzzy soft point
s.f.p.	Soft fuzzy point
$\tilde{X}$	The absolute soft set
$\tilde{\emptyset}$	The null soft set
Iff	If and only if
$\circ$	The composition of functions
AqB	A is quasi coincident with B
$A \not{q} B$	A is not quasi coincident with B
$\tilde{\cap}$	The intersection of two soft sets
$\tilde{\cup}$	The union of two soft sets
$\tilde{\setminus}$	The difference of two soft sets
$\tilde{\in}$	Belonging of two soft sets
$\tilde{\subseteq}$	Soft subset
$P_{\propto}$	TypeI fuzzy point
$P_{\propto}^{x}$	TypeII fuzzy point
$P_{\propto}^{A}$	TypeIII fuzzy point
$P_{\propto}$	TypeI fuzzy soft point
$P_{\propto}^{e}$	TypeII fuzzy soft point
$P_{\propto}^{B}$	TypeIII fuzzy soft point

$\phi(x)$	Structure function
$\mu_{\hat{A}}(x)$	Membership function
$F^{e}{}_{p_{\propto}}^{x}$	TypeI soft fuzzy point
$F_{p_{\propto}}^{x}$	TypeII softfuzzy point
$e_{F}{}_{p_{\propto}}^{x}$	Type IIIsoft fuzzy point
$F^{e}{}_{p_{\propto}}^{F(e)}$	Type IVsoft fuzzy point
$F^{e}{}_{p_{\propto}}^{(A\cap F(e))}$	TypeV soft fuzzy point
Fss	Fuzzy soft set
Sfs	Soft fuzzy set
${}_{f}A_{s}$	Fuzzy soft set A
${}_{s}A_{f}$	Soft fuzzy set A
$f^{\tau}s$	Fuzzy soft topology
$s^{\tau}f$	Soft fuzzy topology
nhd	The neighborhood
f-nhd	Fuzzy neighborhood
s-nhd	Soft neighborhood
pr.	Probability
$\tilde{P}r.$	Fuzzy probability
$\tilde{\tilde{P}}$r.	Soft probability

ε	Uncertain element
a	Uncertain measure
RI	Reliability index
R_i	The reliability of component i
R_s	The reliability of a system
$\tilde{R}_i$	The fuzzy reliability of component i
$\tilde{R}_s$	The fuzzy reliability of a system
$\tilde{\tilde{R}}$	Soft reliability
fR_s	Fuzzy soft reliability
sR_f	Soft fuzzy reliability
IFS	Intuitionistic fuzzy set
$\tilde{A}^i$	Intuitionistic fuzzy set A
P_i	The Path set i
C_i	The Cut set i
Ω	Sample space
FT	Fault tree
FFT	Fuzzy fault tree
SFT	Static fault tree
DFT	Dynamic fault tree
DFFT	Dynamic fuzzy fault tree
ET	Event tree

RBD	Reliability block diagram
	RBD
DRBD	Dynamic Reliability block diagram

List of Tables

List of Figures

Introduction

First of all, we classify the fuzzy sets families which produced seven different families and two new definitions for fuzzy points, they become three different fuzzy points which are in term of definitions of fuzzy neighborhoods for these fuzzy points. It is important to mention that the definitions of the fuzzy adherence point, accumulation fuzzy point, fuzzy interior point, fuzzy exterior point and fuzzy boundary points with studying the relationships between them are presented.

Moreover, a new fuzzy separation axiom are related to three types of fuzzy points and fuzzy compactness are produced, it is considered the basic of mathematics. The same procedures are introduced with the soft points.

In the second part, a definition of families which is called soft sets and by using the composition technique between fuzzy points and soft points we get new three different fuzzy soft points also via nine combinations between soft points and fuzzy points, new five soft fuzzy points are extracted.

For the new three different fuzzy soft points, fuzzy soft topological space also for the new five different soft fuzzy points, soft fuzzy topological space is studied respectively.

It worth to say that it is possible to define adherence, accumulation, exterior, interior and boundary points according to fuzzy soft sets and on the soft fuzzy sets with studying the separation axioms on them added to the above as well.

Soft probability and soft fuzzy reliability in topological spaces are presented in light of concepts of soft sets and probability theory, to show the relation between soft probability and topology is presented via an illustrative example. Evaluation of reliability for different types of systems as an application concern different kind of uncertainty by using distinct mathematical tools as vague sets, intuitionistic fuzzy sets, fuzzy soft sets and soft fuzzy set is presented and added to this applications to clarify the relation between the topology and the confidence interval through an example.

L.Zadeh in 1965 introduced the concept of fuzzy set theory since different problems in real life contains different types of uncertainty and vagueness. He putting the underpinning of the fuzzy set theory , his thoughts concentrated on the idea of degree or grade of membership , which is the concept that became the backbone of fuzzy set theory. In present it is found that many studies noticed by engineers and statisticians that different problems in real life involved different types of uncertainty as vagueness and imprecision on data . Through their published literature for either books or papers in many years ago and for different international publishing houses which are famous in the world they have talked about only two families of fuzzy sets .

In the view of a deep analytical study for these literatures , new five families found ,a classification for the families that involve the two families of which are named by us second family and third family ,it is worth to say that our new five families, are named first, fourth , fifth , sixth and seventh families respectively, they can be worth to care because they were played a necessary role via the several applications , for example , the engineering application in the case of studying the sections of a specified electrical waves , it is meant by a section is the partition of the domain which is studied in the partitions of the domain for the wave as an example for the first family, another example for the fourth family is the studying of different electrical waves on all the partitions of the domain or some domain partitions . Here appear the importance of these new families, also a third example of applications is in fuzzy reliability safety to avoid roads risk is the design of the chassis of vehicles , the metal used in the structure of chassis composed of a steel mixed with manganese have pneumatic holes as a qualitative property which make the chassis more flexible to absorb the vibrated waves which happened especially in bad roads to a void the dangerous failure in vehicles by absorbing the impact in roads as a damped waves . An illustrative examples are presented about the families mentioned above.

The seven families have an influence on topological spaces, they open a wide area especially its' applications in engineering ,control systems , nano , neural networks, total quality management , GIS, economics , medical science , social science , environment , teaching evolution , computer science ,etc, that is in a wide branches of science involves data which are not always all crisp , deterministic and precise in character . Chang's fuzzy topology] arises interest of many researchers to introduce generalized fuzzy closed sets in it The concept of generalized closed sets in topological spaces was introduced by Levine in (1980).. Sostack introduced the notion of smooth topology as an extension of Chang and Lowen's fuzzy topology and developed the theory of smooth topological spaces . After that several researchers assigned to the fuzzy topology in the sense of Chang- Lowen as the topology of fuzzy subsets. Coker and his colleagues introduced the notion of intuitionistic fuzzy topological space. Samanta and Mondial produced the notion of intuitionistic gradation of openness as an extension of smooth topology.

In the current study the first type of fuzzy points is redefined and two new types of fuzzy points are introduced with examples . By virtue of the classifications of fuzzy sets. it can be characterized the fuzzy topology by means of neighborhood (in short nhd) systems, a four new types of nhd's , that help the engineers and scientist for deep study to the graphs or a complete band of graphs on all the definite intervals assigned on the domain universe in order to study its properties and the range of variation happened in it between an interval and the another , which means the answer of the question why we classify the fuzzy sets ?

Many theories are deal with uncertainty problems after Zadehs fuzzy set theory like Pawlak (1982) produced the concept of the rough set theory as a mathematical tool to deal with the problems of uncertainty.

K.T.Atsnassov (1986) introduced the concept of Intuitionistic fuzzy set theory which has many applications in several fields of real world as in reliability theory.In (1993) Vague sets was presented by Gau and Buchrer .Molodtsov (1999) innovative the concept of soft set which is a vividly one and started to developed the basic of the corresponding theory as a new procedure to modeling uncertainties as a general mathematical tool for dealing with uncertain objects, which represents the track of scientific competitive between the countries.

P.K.Maji and his coauthor (2001) they combined fuzzy sets and soft sets via the study of soft set on a family of fuzzy sets called these sets a fuzzy soft sets , in other words it is a soft set defined on the family of fuzzy sets . They presented some properties and relations for these fuzzy soft sets. Maji et al (2003) gave some new definitions on soft sets and they studied the application of soft set theory in the problems of decision making. Many researchers have applied this concept in different problems. Further the fuzzy soft sets takes a wide area of applications in engineering , economic , control systems , decision making etc.

Chang (1968) produced the concept of fuzzy topological spaces which arises interest of many researchers to introduce generalized fuzzy closed sets in it.C.K.Wong (1974) , produced the fuzzy points with some properties .B.Hutton (1975), defined the notion of fuzzy normality in fuzzy topological space and in (1980)

,B.Hutton and I.Reilly introduced the concept of separation axioms in fuzzy topological space .R.H. Warren, (1979) study Fuzzy topologies characterized by neighbourhood systems . Levine 1980, introduced the notion of generalized closed sets (g. closed set) in topological spaces.Since there is much attention has been paid to generalize the basic concepts of classical topology in fuzzy setting and thus modern theory of fuzzy topology has been developed in (1980) p.p. Ming etal introduced the concepts of quasi- coincidence as quasi-neighborhoods .R.Lowen , in 1981 produced compact hausdorff fuzzy topological spaces are topological. A.A.Ramedan in 1992 studied the Smooth topological Spaces .B.M.Juk and Y.E.Jum in 1995 they presented some applications of fuzzy topology .The families of soft sets were defined on P(X) several researchers as N.Cagman , etal in (2001), defined anew family of soft set named $S_2(X)$ .M.shabir and M.Naz in (2011) presented the concept soft topological space and they defined some basic concepts concerning their study. In (2011), S.Hussain and B.Ahmed they proved several properties of soft open sets and soft closed sets and defined soft interior . In (2012) S:S.Atmagen M.Akdag, I . Zorolutana and W.K.Min[44] defined another new type of soft families of soft set . G.Xuenchong . (2015) presented a new family type of soft sets named central soft set . Luay A.A.H. Zahraa M.N in 2017 they presented e-Soft Separation Axioms in Soft Topological Space.

N.Cagman, Serkan Karatas , Serdar Engingllu used the concept of the soft sets to produce the concept of the soft topological space and its' properties .

In our study we reformulate some definitions in view of the mathematical analysis concepts this impose that the definitions must be become more precise and simple. The combination between the soft sets and fuzzy sets to find a fuzzy set such that its membership function is a combination of the fuzzy function with the soft function , say f is a function from $\mathcal{P}(X)$ to $I = [0,1]$ and F is a function from E to $\mathcal{P}$(X) where X is a universal set , E is the set of parameters of the universal set X . We classify the families that contains these sets which are represent the combination of soft families with fuzzy families to introduce a fuzzy families which depends on them .

introduced four families of soft sets denoted by $S_1(X), S_2(X), S_3(X)\ and\ S_4(X)$ with some basic notions , Three different soft points mentioned in [14] which are related with the first soft family $S_1(X)$.

It is importent to mention that in our study new different fuzzy soft points , produced via the composition between the three different types of the fuzzy points which produced by [6] with the different three types of soft points mentioned in [14] .Illustrative examples are appended . Shabir and Naz introduced soft topology by using soft sets with some basic properties, after that many researchers improved this study and gave new results [29]. Tanay and Kandemir defined fuzzy soft topology on a fuzzy soft set and gave an introductory theoretical base to carry further study on this topic .

Several researchers introduced new classes of separation axioms and gave studied of their properties[55] . A new compactness spaces called Gem-compact space , points space under the idea of Gem-set in[59] was presented.

In the present study an analytical study is presented for the composition of the fuzzy topological spaces with soft topological spaces and we get a fuzzy soft topological spaces , moreover we redefined the axioms of separation by using the new fuzzy soft points listed in table (2-1), base , sub base for these spaces and fuzzy soft compactness with illustrative examples.

Also an analytical study is produced via the combination of the soft topological spaces with fuzzy topological spaces and we obtain a soft fuzzy topological spaces , moreover the axioms of separation by using the new soft fuzzy points listed in table (3-1) , base , sub base for these spaces and soft fuzzy compactness are presented.

As an application, the fuzzy soft set is used in totally quality development for active teaching.

The vague sets are used to analysis the reliability of systems like weapon systems which are expensive or dangerous to measure experimentally, expert thoughts are used to provide the fault information but estimate's usually is uncertain and it cannot rule out any possibility of system failures including power systems, nature reasons, human error and others, see[156].

Hence by vague sets it can solve this kind of problems. For more details see[68].

Sandler (1963) presented different applications in reliability especially the design military weapons and become more applications in industry and studied the calculation of reliability for the series system and for the parallel system without maintenance . [163]

Before many years, reliability was vague term to some extent in spite of it is a probability and it was not of concern in designing, developing and making decision. Reliability become more connected with all main stages of evaluating in economic, military industries , nuclear power plant and aerospace systems on the all levels of the design in operation research and maintenance. [141]

Since the fault tree analysis (FTA) is one of important methods of calculating the reliability of complex system this research is concerned with the Fault Tree analysis technique and its relationship with other techniques

The first research on fuzzy fault tree analysis was done by Tanaka et al in (1983) the treated probabilities of basic event as trapezoidal fuzzy numbers.

To simply the calculation Cheng and Mon (1993) considered the failure probability of basic events as triangular fuzzy numbers.

Al – Ali and Subrie , (2002) used a path set method to calculate the reliability of a power plant which is considered as a complex system .

The concept of Intuitionistic Fuzzy Set (IFS) was permitted to incorporate simultaneously the membership degree and the non-membership degree of each element. Fuzzy logic is becoming more popular resulting in many applications in several fields of real world including reliability theory, since it is not possible having accurate and complete information about the system. Therefore, in many cases it is very difficult to calculate the system reliability. To handle the complete information, the approach of fuzzy set theory can be used to estimate the system reliability the idea of Fuzzy reliability has been proposed and developed by several authors D.Pandey and L.A. Zadeh based on possibility assumption or fuzzy state assumption. Recently the researchers L.A. Zadeh proposed that the reliability of every component is a fuzzy variable and evaluated system reliability.
Vague sets were used in reliability system analysis. More over fuzzy soft set concept which presented by Maji et al in (2001) is used to calculate the reliability of different models of systems . Wide area of application is for fault tree of different types (static and dynamic)[144] are used to solve the problems of reliability via the quantative and qualitative analysis by applying the Boolean algebra [119].

In our study we define the soft generalized vague set and also the operations equality , subset, null set union , Intersection OR,AND operator on it with applications

we present fuzzy probability, soft probability, probabilistic soft sets and the relationship between fuzzy reliability and classical reliability. Soft fuzzy reliability in topological spaces is presented in view of soft approximation and by considering reliability as probability according to its definition.

In addition to that, the classical reliability is calculated in view of Lui concept of uncertain measure for it, and an applications in different types of uncertainty through illustrative examples in medical diagnoses and in an active teaching related with total quality, also in calculating reliability for many models of systems by using different mathematical tools as fuzzy sets , vague sets , intuitionistic fuzzy sets, type-2 fuzzy set, soft intuitionistic fuzzy sets more over several techniques as fault tree , event tree decomposition tree, dynamic fault tree , triangular fuzzy numbers and converting confidence interval to a triangular fuzzy numbers with illustrative examples.

Chapter One

Fuzzy Sets and Soft Sets

1. Preliminaries:

In this chapter definitions about fuzzy sets have given and a new five families of fuzzy sets was introduced and added to the two families which introduced by [2,4,6,8] . After analytical study a classification was done for the families , also introduced new types of nhds and quasi nhds. We redefined the first type of fuzzy points presented in [13] and introduced two new types of fuzzy points. with some notes . The investigation of basic operations on fuzzy sets are presented via illustrative examples. Some properties for limit point ,adherence point, interior , exterior and boundary points are discussed . Also for the soft points mentioned in [34] are discussed

A fuzzy topology is presented and base ,sub base , fuzzy separation axioms and fuzzy compact related with the types of fuzzy points also a soft topology with base, sub base and soft separation axioms and soft compact related with the different types of soft points are produced .

1.1. Some Definitions and Concepts

In this section definitions of fuzzy sets, t-norms , t-co norms and its types with fuzzy operations as union , intersection and complements with some theorems also types of fuzzy points are presented.

Definition 1.1.1:

If X is a collection of objects denoted generically by x , then a fuzzy set $\tilde{A}$ in X is a set of ordered pairs

$$\tilde{A} = \{(x, f_A(x)) : x \in X\}$$

$f_A(x)$ is called the membership function or grade of membership (also degree of compatibility or degree of truth)of x in $\tilde{A}$.

Which assign a real number $f_A(x)$ in the interval [0,1] , to each element $x \in X$. where the value of $f_A(x)$ at x shows the grade of membership of x in A.

i. $\tilde{0} = \{(x, 0),\ \forall x \in X \times I\}$ is the null fuzzy set.
ii. $\tilde{1} = \{(x, 1),\ \forall x \in X \times I\}$ is the absolute fuzzy set.

Definition 1.1.2 : A t-norm operation denoted as t(x,y) is a function form $[0,1] \times [0,1] to [0,1]$ that satisfies the following condition $\forall\ w, x, y, z \in [0,1]$

1. t(0,0) = 0 , t(x,1) = t(1,x) = x
2. $t(x, y) \leq t(z, w)\ if\ x \leq z\ and\ y \leq w$ (monotonicity)
3. $t(x, y) = t(y, x)$ (commutativity)
4. $t(x, t(y, z)) = t(t(x, y), z)$ (associativity)

Definition1.1.3: A t-conorm operation denoted as s(x,y) is a function from $[0,1] \times [0,1] to [0,1]$ that satisfies the following condition $\forall\ w, x, y, z \in [0,1]$

1. s(1,1) = 1 , s(x,0) = s(0,x) = x
2. $s(x, y) \leq s(z, w)\ if\ x \leq z\ and\ y \leq w$ (monotonicity)
3. $s(x, y) = s(y, x)$ (commutativity)
4. $s(x, s(y, z)) = s(s(x, y), z)$ (associativity)

Definition1.1.4: Let A be a fuzzy set on the universal set X then the complement of $\tilde{A}\{\ in\ short\ c\tilde{A}$ which is defined as a function $c: [0,1] \rightarrow [0,1]$ which assigns a value c(f_A(x)) to each membership grade f_A(x) for any given fuzzy set $\tilde{A}$.

To produce meaningful fuzzy complements function c must satisfy at least the following 2 axiomatic requirements:

Axiom 1: c(0) = 1 and c(1) = 0 (boundary condition)

Axiom 2: $\forall\ a, b\ \in [0,1]\ if\ \ a \leq b\ then\ c(a) \geq c(b)$ (monotonicity)

Let a complement cA defined by a function **c:[0,1]→[0,1]** which assigns a value c($f_A(x)$) to each membership grade $f_A(x)$ of any given fuzzy set $\tilde{A}$.
Function c is required:
(1) for Axiom to produce correct complement for crisp set
(2) according to Axiom 2 to be monotonic decreasing when a membership grade in $\tilde{A}$ increases (by changing x) , the corresponding membership grade $c\tilde{A}$ must not increase as well , it may decrease or at least remain the same.

Note 1.1.5

i- There are many functions that satisfy both Axioms 1 and 2

ii- For any particular fuzzy set $\tilde{A}$, different fuzzy sets $c\tilde{A}$ can be said to constitute its complement each being produced by distinct function c.

iii- All functions that satisfy the axioms form the most general class of Fuzzy complements

iv- (Axioms 1 and 2) be called the axiomatic skeleton for fuzzy complements .

It is desirable to consider various additional requirements for fuzzy complements. Each of the reduces the general class of fuzzy complements to a special subclass.

Two of the most desirable requirements are the following .

Axiom 3: c is a continuous function

Axiom 4: c is involutive , which means c(c(a))=a ∀

In turns out the four axioms are not independent as shown in the following

Theorem 1.1.6 :[5] Let a function: c:[0,1]→[0,1] satisfy Axioms **2** and **4** , Then **c** also satisfies Axioms **1** and **3** , moreover **c** must be a bijective function.

It follows from theorem above that all involutive complements from a special subclass of all continuous complements , which in turn forms a special subclass of all fuzzy complements.

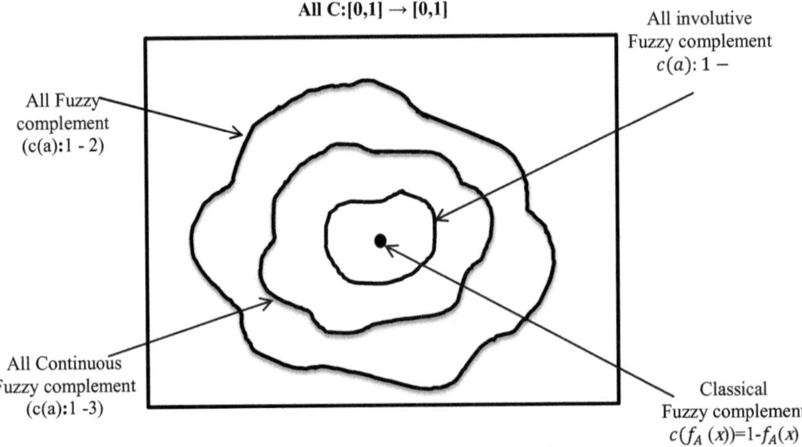

Figure (1-1). shows structure of the basic classes of fuzzy complements

Examples of some types of t-norm and t-conorm that are frequently used as fuzzy intersection , fuzzy union and complement(each defined $\forall\, a, b \in [0,1]$) . The following t- norms and t-conorms are dual with respect to the standard fuzzy complement as in the table below

Table (1-1) t- norms and t-co norms (or s-norm)

The types	t-norms	t-conorms	Complement
Standard	i(a,b) = min(a,b)	$u(a,b) = \max(a,b)$	C(a)=1-a
Algebraic	i(a,b) = ab	$u(a,b) = a + b - ab$	C(a)=1-a
Bounded	i(a,b) = max(0,a+b-1)	$u(a,b) = \min(1, a + b)$	C(a)=1-a
Drastic	$i(a,b) = \begin{cases} a & when\ b = 1 \\ b & when\ a = 1 \\ 0 & otherwise \end{cases}$	$u(a,b) = \begin{cases} a & when\ b = 0 \\ b & when\ a = 0 \\ 1 & otherwise \end{cases}$	C(a)=1-a

The relation between the four fuzzy unions via their graphs is :

Max (a ,b)$\leq a + b - ab \leq \min(1, a + b) \leq \mu_{max}(a,b)$

$\forall\, a, b \in [0,1], where\ \mu_{max}\ denotes\ the\ drastic\ union.$ And

$i_{min}(a,b) \leq \max(0, \mathrm{a} + \mathrm{b} - 1) \leq \min(\mathrm{a}, \mathrm{b})$

$\forall\, a, b \in [0,1]$) where i_{min} denotes the drastic intersection.

Now we will include some theorems and definitions by which we extract both t-norm , t-conorm and complement

Definition 1.1.6$_1$

We will call the function g : $[0,1] \to R$ s.t. g(0)=0 is

1- increasing generator if g is continuous and strictly increasing

2- decreasing generator if g continuous and strictly decreasing g(1)=0

3- the pseudo inverse of increasing generator g denoted by $g^{(-1)}$ is a function from R to [0,1] defined by

4- $$g^{(-1)}(a)=\begin{cases}0 & for\ a\in(-\infty,0)\\ g^{-1} & for\ a\in[0,g(1)]\\ 1 & for\ a\in(g(1),\infty)\end{cases}$$

Where g^{-1} is the ordinary inverse of the function g

5- The pseudo inverse of decreasing generator g denoted by $g^{(-1)}$ is a function from R to [0,1] defined by

$$g^{(-1)}(a)=\begin{cases}1 & for\ a\in(-\infty,0)\\ g^{-1}(a) & for\ a\in[0,g(0)]\\ 0 & for\ a\ (g(0),\infty)\end{cases}$$

Theorem 1.1.6_2

1- **If** c is a fuzzy complement , then there exist a continuous function g from [0,1] to R s.t. g(0)=0, g is strictly increasing and

$C(a) = g^{-1}(g(1) - g(a))$ $\forall a \in [0,1]$

2- **If $C(a) = g^{-1}(g(1) - g(a))$ $\forall a \in[0,1]$** and g from [0,1] to R s.t. g is strictly increasing , then c is fuzzy complement satisfy the axioms c_1 to c_4 .

3- **Let u and i are the binary operations on the unit interval [0,1] then,**

U is an Archimedean t-conorm (i t-norm) iff there exist increasing generator g s.t.

$$U(a,b) = g^{(-1)} (g(a) + g(b))$$

$$i(a,b) = g^{(-1)}(g(a) + g(b) - g(1)) \ \forall\ a,b \in [0,1]$$

Example 1.1.6_3

Let $g(a) = a^w$, $w > 0$ find the complement , t-norm and t-conorm

Solution . g(0)=0 , g(1) = 1 and $g^{-1}(a)=a^{1/w}$

1- $C(a) = g^{-1}(g(1) - g(a)) = g^{-1}(1 - a) = (1-a)^{1/w}$

2- $U(a, b) = \min(1, (a^w + b^w)^{1/w})$

3- $i(a, b) = \min(1, (a^w + b^w - 1)^{1/w})$

Among the big verity of fuzzy intersection, union and complements ,the standard fuzzy operations have certain properties that give them a special significance , since they are the only operation that satisfy the cut worthy strong worthy properties . The standard fuzzy operations occupy specific positions in the whole spectrum of fuzzy operations

Definition 1.1.7:

Let $\tilde{A} = \{(x, f_A(x)): x \in X\}$ and $\tilde{B} = \{(x, f_B(x)): x \in X\}$ be two fuzzy sets in X. Then their Union, intersection and complement are also sets with the membership functions defined as follows:

(i) $f_{A\cup B}(x) = \max\{f_A(x), f_B(x)\}, \forall x \in X$

(ii) $f_{A\cap B}(x) = \min\{f_A(x), f_B(x)\}, \forall x \in X$

(iii) $[f_A(x)]^c = 1 - f_A(x), \forall x \in X$

Theorem 1.1.8

Let $\tilde{A}, \tilde{B}$ and $\tilde{C}$ be fuzzy sets in a set X, the fllowing statement hold:

a) (Commutatively): $\tilde{A} \vee \tilde{B} = \tilde{B} \vee \tilde{A}; \tilde{A} \wedge \tilde{B} = \tilde{B} \wedge \tilde{A}$.
b) (Associatively): $(\tilde{A} \vee \tilde{B}) \vee \tilde{C} = \tilde{A} \vee (\tilde{B} \vee \tilde{C}); (\tilde{B} \wedge \tilde{A}) \wedge \tilde{C} = \tilde{A} \wedge (\tilde{B} \wedge \tilde{C})$.
c) (Idempotency): $\tilde{A} \vee \tilde{A} = \tilde{A}; \tilde{A} \wedge \tilde{A} = \tilde{A}$.
d) (Distributivity): $\tilde{A} \vee (\tilde{B} \wedge \tilde{C}) = (\tilde{A} \vee \tilde{B}) \wedge (\tilde{A} \vee \tilde{C}); \tilde{A} \wedge (\tilde{B} \vee \tilde{C}) = (\tilde{A} \wedge \tilde{B}) \vee (\tilde{A} \wedge \tilde{C})$.
e) (Absorption): $\tilde{A} \vee o_X = \tilde{A}; \tilde{A} \wedge 1_X = \tilde{A}$.
f) (De Morgan's): $(\tilde{A} \vee \tilde{B})^c = \tilde{A}^c \wedge \tilde{B}^c, (\tilde{A} \wedge \tilde{B})^c = \tilde{A}^c \vee \tilde{B}^c$.
g) (Involution): $(\tilde{A}^c)^c = \tilde{A}$.
h) (Equivalence formula): $(\tilde{A}^c \vee \tilde{B}) \wedge (\tilde{A} \vee \tilde{B}^c) = (\tilde{A}^c \wedge \tilde{B}^c) \vee (\tilde{A} \wedge \tilde{B})$.

i) (Symmetrical difference formula): $(\tilde{A}^c \wedge \tilde{B}) \vee (\tilde{A} \wedge \tilde{B}^c) = (\tilde{A}^c \vee \tilde{B}^c) \wedge (\tilde{A} \vee \tilde{B})$.
j) (Difference): $(\tilde{A} \backslash \tilde{B}) = \tilde{A} \wedge \tilde{B}^c$.

Remark(1.1.9):
Let $\tilde{A}$ and $\tilde{B}$ be fuzzy sets in a set X, then $\tilde{A} \wedge \tilde{B} \leq \tilde{A}$ and $\tilde{A} \wedge \tilde{B} \leq \tilde{B}$.

Remark(1.1.10)
The excluded-middle law is no longer true: $\tilde{A} \wedge \tilde{A}^c \neq o_X, \tilde{A} \vee \tilde{A}^c \neq 1_X$.

Definition (1.1.11)
For any family $\{\tilde{A}_j : j \in J\}$ of fuzzy sets in X, the union $\vee_{j\in J}\tilde{A}_j$ and the intersection $\wedge_{j\in J}\tilde{A}_j$ are defined by:

$$\vee_{j\in J}\tilde{A}_j = \{(x, sup\{\tilde{A}_j : j \in J\}), \forall x \in X\},$$
$$\wedge_{j\in J}\tilde{A}_j = \{(x, inf\{\tilde{A}_j : j \in J\}), \forall x \in X\}.$$

Theorem (1.1.12)
For fuzzy subsets $\tilde{A}, \tilde{A}_j, j \in J$ of a set X

a) $\tilde{A} \wedge (\vee_{j\in J}\tilde{A}_j) = \vee_{j\in J}(\tilde{A} \wedge \tilde{A}_j)$
b) $\tilde{A} \vee (\wedge_{j\in J}\tilde{A}_j) = \wedge_{j\in J}(\tilde{A} \vee \tilde{A}_j)$
c) $(\vee_{j\in J}\tilde{A}_j)^c = \wedge_{j\in J}\tilde{A}_j^{\,c}$
d) $(\wedge_{j\in J}\tilde{A}_j)^c = \vee_{j\in J}\tilde{A}_j^{\,c}$

Definition 1.1.13 :

The support of the fuzzy set $\tilde{A}$ which is denoted by supp($\tilde{A}$) is the set of the element that have non zero degrees of membership in $\tilde{A}$ that is :

$\mathrm{supp}(\tilde{A}) = \{\mathrm{u} : \mathrm{u} \in \mathrm{X} \text{ and } \mathrm{f_A}(\mathrm{u}) > 0\}$and $\tilde{A}$ is called normal, if $\exists$ u $\in$ X , $\ni$ $\mathrm{f_A}(\mathrm{u}) = 1$.

Types of fuzzy points 1.1.14:
Now we redefined the first type of fuzzy points and introduced two new fuzzy points as in below :

1. **Type I :** For some $x \in X$ $and\ o < \alpha \le 1$, we say that P_α^x is classical fuzzy point on the fuzzy point of type I
if $P_x^\alpha(y) = \begin{cases} \alpha & if\ y = x \\ 0 & if\ y \neq x \end{cases}$

Example 1.1.15: Let X={x_1,x_2,x_3} and $\alpha = 0.131$

$$P_{x_2}^\alpha = \{(x_1, 0), (x_2, 0.131), (x_3, 0)\}$$

2. **Type Π :** Let A⊂X and for some 0<$\alpha \le 1$. The formula P_α^A is called the fuzzy point of Type II,
where $P_\alpha^A(y) = \begin{cases} P_y^\alpha & if\ y \in A \\ 0 & if\ y \notin A \end{cases}$

Example 1.1.16 : Let X={x_1,x_2,x_3} , A={x_2,x_3}, $\alpha = 1$

$$P_1^A = \{(x_1, 0), (x_2, 1), (x_3, 1)\}$$

Noted that $P_\alpha^A = \cup_{x \in A} P_x^\alpha$

3. **Type ш :** For any 0<α<1,the formula $p_\alpha$is called the fuzzy point of type ш
such that $P_\alpha(y) = \alpha\ \forall\ y \in X$

Example 1.1.17: Let X={x_1,x_2,x_3} , $\alpha = 0.23$

$$P_\alpha = P_{0.23}=\{(x_1, 0.23), (x_2, 0.23), (x_3, 0.23)\}$$

Noted $that\ P_\alpha = \cup_{x \in A} P_x^\alpha$ also we can denoted the fuzzy point of type ш by P_α^X.

Definition (1.1.18)

A fuzzy point p_x^λ in X is said to be q-coincident with a fuzzy set $\tilde{A}$ in X, denoted by $p_x^\lambda qA$ iff $\lambda + f_A(x) > 1$.

If $f_A(x) + g_B(x) > 1$ for some $x \in X$ then $\tilde{A}$ is called q-coincident with $\tilde{B}$. Otherwise if $f_A(x) + g_B(x) \le 1$, for every $x \in X$, then $\tilde{A}$ is called not q-coincident with $\tilde{B}$ and it is denoted by $\tilde{A}$q $\tilde{B}$.

Example 1.1.19:

Let X={a,b,c} be a set and $p_{\alpha}^{0.5}$ be a fuzzy point in X , and $\tilde{A}, \tilde{B}$ and $\tilde{C}$ are fuzzy sets in X defined as below:

f_A(a)=0.1 , g_B(a)=0.9 , h_C(a)=0.8

f_A(b)=0.6 , g_B(b)=0.3 , h_C(b)=0.4

f_A(c)=0.5 , g_B(c)=0.4 , h_C(c)=0.6

$p_{\alpha}^{0.5}$is q-coincident with a fuzzy set $\tilde{B}$ since $0.5 + g_B(\text{a}) = 0.5 + 0.9 = 1.4$

Lemma 1.1.20 : For any $\tilde{A} \in I^X$, the following hold

1. If $\tilde{A} \neq 0$, then $\exists x \in X$ and $0 < \lambda \leq 1$ such that $p_x^{\lambda} q\tilde{A}$.
2. If $\tilde{A} \neq 0$, then $\exists x \in X$ and $0 < \alpha \leq 1$ such that $p_x^{\alpha} q\tilde{A}$.
3.

Proof: 1) Suppose that $\tilde{A} \neq 0$, then $\exists x \in X$ such that $f_A(x) \neq 0$. Then we can take real positive number λ such that $\lambda + f_A(x) > 1$. So, $p_x^{\lambda} q\tilde{A}$.

2) Suppose that $\tilde{A} \neq 0$

Case I. If there is only one point $x \in X$ such that $f_A(x) \neq 0$, so we can take $0 < \alpha \leq 1, \alpha + f_A(x) \leq 1$. But $\forall y \in X \ni x \neq y, f_A(y) = 0$, then $\alpha + f_A(y) \leq 1$. Therefore $p_x^{\lambda} q\tilde{A}$.

Case II. If there is $Y \subseteq X$ such that $f_A(y) \neq 0, \forall y \in Y$ we can also take positive real numbers $0 < \alpha_y < 1$ such that $f_A(y) + \alpha_y < 1, \forall y \in Y$.

Put $\alpha = inf\{\alpha_y\}_{y \in Y}$, thus $f_A(y) + \alpha_y \leq 1, \forall y \in Y$. But $f_A(w)$ for $w \notin Y$. Therefore $f_A(z) + \alpha_y \leq 1, \forall z \in X$. Hence, $p_x^{\alpha} q\tilde{A}$

Definition 1.1.21:

A fuzzy set $\tilde{A}$ is said to be quasi coincident with a fuzzy set $\tilde{B}$ and is denoted by $\tilde{A} q \tilde{B}\ iff\ \exists x \in X \ni f_A(x) + g_B(x) > 1$

Return to example (1.1.19) : f_A(c) + g_B(c) =0.5 +0.6 = 1.1 >1 so $\tilde{A} q \tilde{C}$,but f_A(c) , g_B(c) are not q-coincident since $f_A(c) + g_B(c) = 0.5 + 0.4 = 0.9$

Remark(1.1.22):

Let $p_x^{\lambda} \in FP(X)$ and $\tilde{A}$ be a fuzzy subset of a set X. Then

a) $p_x^\lambda q\tilde{A}$ iff $\lambda > f_{\tilde{A}^c}(x)$
b) $p_x^\lambda \not{q}\tilde{A}$ iff $\lambda \leq f_{\tilde{A}^c}(x)$.

Lemma(1.1.23)

Let $\tilde{A}, \tilde{B}$ and $\tilde{C}$ be fuzzy sets in a set $X, \{\tilde{A}_j : j \in J\}$ be a family of fuzzy sets in X and p_x^λ be a fuzzy point in X. Then

a) If $\tilde{A} \wedge \tilde{B} = \tilde{0}_X$, then $\tilde{A}\not{q}\tilde{B}$.
b) $\tilde{A}\not{q}\tilde{B}$ iff $\tilde{A} \leq \tilde{B}^c$.
c) $\tilde{A}q\tilde{A}^c$.
d) $\tilde{A}\not{q}\tilde{B}$ and $\tilde{C} \leq \tilde{B}$ then $\tilde{A}q\tilde{C}$.
e) $\tilde{A} \leq \tilde{B}$iff $p_x^\lambda q\tilde{B}$ for every $p_x^\lambda q\tilde{A}$.
f) $p_x^\lambda q(\vee_{j\in J}\tilde{A}_j)$ iff there is $j \in J$ such that $p_x^\lambda q\tilde{A}_j$.
g) $p_x^\lambda q(\wedge_{j\in J}\tilde{A}_j)$ then $p_x^\lambda q\tilde{A}_j$ for all $j \in J$.

Lemma(1.1.24)

Let p_x^λ be a fuzzy point in X and $\tilde{A}$ be any fuzzy subset of X. Then the following statements are equivalent:

a) $p_x^\lambda \in \tilde{A}$
b) $p_x^\lambda \not{q}\tilde{A}^c$
c) $p_x^{1-\lambda} q\tilde{A}$

Proof.

$(a) \Leftrightarrow (b)$ direct .

$(a) \Leftrightarrow (c)$ *Since* $p_x^\lambda \in \tilde{A} \Leftrightarrow \lambda \leq f_A(x)$

So we have that $\Leftrightarrow 1-\lambda > 1 - f_A(x)$

$= [f_A(x)]^c$

$\Leftrightarrow p_x^{1-\lambda} q\tilde{A}$

1.2.Families of Fuzzy Sets:

In this section we will give some definitions and obtain families of fuzzy sets which are worth to care. According to the current analytical study for fuzzy sets a classification of fuzzy sets is presented for the families and we defined and denoted these families . Basic operations on fuzzy sets are investigated via an illustrative examples. Notes are listed about the relation between the families.

We introduce the following definitions to classify the families of fuzzy sets

:

Definition 1.2.1.:

I. Let μ (X , I)= $\{f;\ f: X \to I\}$ be the set of all functions f on the universal set X to the unit interval I=[0,1]

II. Let $\phi_A(X) = \{f_{iA} : \forall f_i \in \mu(X, I)\}$, $f_{iA}(x) = f_i(x)$, $\forall x \in A$ be the set of all restriction functions in μ(X ,I) with respect to the subset A of X

.

III. Let $\Psi_{Ai}(X) = \{f_{j_{A_i}} : \forall f_j \in \mu(X, I)\} \forall A_i \subseteq X$, where $f_{A_i}(x) = f_i(x)$, $\forall x \in A_i$ be the set of all restriction functions in μ(X,I) with respect to the subsets A_i of X

First family:

We introduce a new family, which it means that it is containing all fuzzy sets that depends on all sets contained in the power set of $X(\forall A \in 2^X)$ for fixed function, f *such that* $f: X \to I$, we call it the first family and denoted it *by* I_1^X , that is :

$$I_1^X = \{\tilde{A}_i ; \tilde{A}_i = (x, f_{A_i}(x)) : \forall A_i \in 2^X , f_{A_i}(x) = f(x) : \forall x \in A_i \}$$

Example 1.2.2:

Let X={1,2,………,100}

And $f(x) = \begin{cases} \frac{1}{x} & \forall x \in X/\{100\} \\ 0 & x = 100 \end{cases}$

for $\forall A \in \boldsymbol{P}(X)$, $\tilde{A} = \{(x, f_A(x)) : f_A(x) = f(x), \ \forall x \in A\}$

For A={1} , $\forall x \in X/\{1\}$ undefined

$\tilde{A} = \{(1,1)\}$,

B={1, 2 , 99}, $\tilde{B} = \{(1,1), (2,\frac{1}{2}), (99,\frac{1}{99})\}$

Note 1.2.3 : In [9],[18] the fuzzy subset A of a fuzzy set B is defined as follows :

if $\tilde{A} \leq \tilde{B}$ then $f_A(x) \leq f_B(x) \quad \forall x \in X$

In our study , $\tilde{A} \subseteq \tilde{B} \leftrightarrow f(x) \leq f(x) \quad \forall x \in X$

Definition 1.2.4: The Subset for fuzzy sets in I_1^X is defined as follows :

$\forall \tilde{A}, \tilde{B} \in I_1^X$, we call that the fuzzy set $\tilde{A}$ *is a* fuzzy subset to the fuzzy set $\tilde{B}$

($\tilde{A} \leq \tilde{B}$) if it is satisfy the following *: 1-* A⊆ B *2-* $f_A(x) \leq f_B(x), \forall x \in A$ where

$\tilde{A}=\{(x, f_A(x)) \mid x \in A\}$

$\tilde{B}=\{(x, f_B(x)) \mid x \in B\}$

Definition 1.2.5: The standard Intersection and Union of two fuzzy sets **:**

Let $\tilde{A}, \tilde{B} \in I_1^X$, *then*

$i)\ \tilde{A} \wedge \tilde{B} = \{ (x, \min_{x \in A \cap B} (f_A(x), f_B(x))), \ x \in A \cap B\}$

$ii)\tilde{A} \vee \tilde{B} = \{(x, \max_{x \in A \cap B} (f_A(x), f_B(x))), x \in A \cap B\}$ *and in the case*

$x \in X/A \cap B$ *it is not valid*

From above we conclude that the intersection and union of two fuzzy sets valid within the common points between the domains of the fuzzy sets i.e. the first coordinate in which their functions in this points are defined on both sets .

In example 1.2.2 in I_1^X we illustrate that the set $\tilde{A}$ contains $\{1\}$ which means that its function defined only in the point 1 and undefined in other points $\{2,3,\ldots,100\}$ whereas the function of set $\tilde{B}$ is defined in $\{1,2,99\}$ and undefined in other points on X, so the two operations intersection and union valid only in the common points between the domains of both fuzzy sets , and these compel us to extend the fuzzy sets (i.e. universal set X in order that all the functions of fuzzy sets become defined).

Second family :

The Second family which we denoted it by I_2^X at the subset A of X .

In other words , we fixed the subset A and compute the restriction for all functions $\boldsymbol{f}$ in $\mu(X, I)$,that is

$I_2^X = \left\{\tilde{A}: \tilde{A} = \{(x, f_{iA}(x)), \forall f_i \in \Phi_A(X), f_{iA}(x) = f_i(x), \forall x \in A\}\right\}$

Example 1.2.6 :

Let X= {0,1,2,3,4,5} , and $f(x) = \frac{1}{1+x}$ $\forall x \in X$

$f(0) = 1, f(1)=\frac{1}{2}$, $f(2)=\frac{1}{3}$, $f(3)=\frac{1}{4}$, $f(4)=\frac{1}{5}$, $f(5)=\frac{1}{6}$

And let $g(x) = \frac{1}{3+2x}$, $\forall x \in X$

Let A={0,1,2} ⊂X

$\tilde{B}=\{(x , f_A(x)); x\in A\}=\{(0,1) ,(1,\frac{1}{2}) , (2,\frac{1}{3})\}$

$\tilde{C}=\{(x, g_A(x)) \mid x\in A\} = \{ (0,\frac{1}{3}),(1,\frac{1}{5}), (2,\frac{1}{7})\}$

Definition 1.2.7: The subset for a fuzzy set is defined as follows:

$\forall$ Ã, $\tilde{B} \in I_2^X$

Ã$\leq \tilde{B} \leftrightarrow f_A(x)\leq g_A(x), \forall x \in A$, where

$\tilde{A} = \{(x,\ f_A(x)) , \forall\ x \in A\} , \tilde{B} = \{(x, g_A(x)) , \forall\ x \in A\}$

Definition 1.2.8: The standard Intersection and Union of two fuzzy sets :

Let $\tilde{A} , \tilde{B} \in I_2^X$, *then*

i) $\tilde{A} \wedge \tilde{B} = \{(x , \min_{x\in A}(f_A(x), g_A(x))) , \forall x \in A\}$

ii) $\tilde{A} \vee \tilde{B} = \{(x , \max_{x \in A}(f_A(x), g_A(x))) ; \forall x \in A\}$

Third Family

It means that , it is containing all fuzzy sets which depends on all functions $f_i \in$ μ(X,I) , we denoted it by I_3^X , That $I_3^X = \{$Ã : Ã $= (x, f_i(x)), \forall f_i \in \mu(X, I)\}$

Example 1.2.9:

Let X={1,2,3,4,5}

And $f(x) = \frac{1}{x}$, $\forall\ x \in X$

$g(x) = \frac{1}{1+x}$, $\forall\ x \in X$

$h(x) = \frac{1}{|2-x|}$, $\forall\ x \in X$

and let Ã $= \{(x, f(x)) , \forall\ x \in X\}$, $\tilde{B} = \{(x, g(x)) , \forall\ x \in X \}$

$\tilde{C} = \{(x, h(x)) , \forall\ x \in X\}$, then

Ã $= \{(1,1), \left(2,\frac{1}{2}\right), \left(3,\frac{1}{3}\right), \left(4,\frac{1}{4}\right), \left(5,\frac{1}{5}\right)\}$

$\tilde{B}=\{(1,\frac{1}{2}), (2,\frac{1}{3}),(3,\frac{1}{4}),(4,\frac{1}{5}),(5,\frac{1}{6})\}$

$\tilde{C}$ ={(1,1),(3,1),(4,$\frac{1}{2}$),(5,$\frac{1}{3}$)}

Definition 1.2.10: The subset for a fuzzy set is defined as follows:

$\forall\ \tilde{A}, \tilde{B} \in I_3^X$

where $\tilde{A}=\{(x, f(x))\ ,\ x \in X\}$

$\tilde{B} = \{(x, g(x))\ ,\ x \in X\}$

$\tilde{A} \subseteq \tilde{B} \leftrightarrow f(x) \le g(x) \qquad \forall x \in X$

Definition 1.2.11 :The intersection and Union of two fuzzy sets :

Let $\tilde{A}, \tilde{B} \in I_3^X,$ *then*

$i)\ \tilde{A} \wedge \tilde{B} = \{(x, \min_{x\in X}(f(x), g(x))), \forall x \in X\}$

$ii)\tilde{A} \vee \tilde{B} = \{(x\ , \max_{x \in X}\ (f(x), g(x))), \forall x \in X\}$

Fourth Family :

It is containing all fuzzy sets depends on all functions

$f_{i_{A_i}} \in \Psi_{Ai}$ (x) and we donate this family by I_4^X , that is

$$I_4^X = \left\{\tilde{A}_i : \tilde{A}_i = \left\{(x_i\ , f_{i_{A_i}}(x)); f_{i_{A_i}} \in \Psi_{Ai}\ (X)\right\}\right\}$$

Example 1.2.12:

Let $X=\{x_1, x_2, \ldots, x_9\}$, $i = 1,2, \ldots ,9$

and $A=\{x_2\ , x_4\ , x_6\ , x_8\}$

$B=\{x_1\ , x_3\ , x_5\ , x_7\ , x_9\}$

$f_{(x_i)} = \frac{1}{(1+i)^2-1}$, $g_{(x_i)} = \frac{1}{2i^2-1}$

$\tilde{A} = \{(x, f(x)),\ \forall x \in A\}$

$=\{(x_2\ ,\frac{1}{8})\ , (x_4\ ,\frac{1}{24})\ , (x_6\ ,\frac{1}{48})\ , (x_8\ ,\frac{1}{80})\}$

$\tilde{B}=\{(x, g_B(x))\ , \forall x \in B\}$

$=\{(x_1,1), (x_3, \frac{1}{17})\ , (x_5, \frac{1}{49}), (x_7, \frac{1}{97}), (x_9, \frac{1}{161})\}$

Definition 1.2.13: The Subset for a fuzzy set is defined as follows:

$\forall\tilde{A}\ , \tilde{B} \in I_4^X$ where $\tilde{A} = \{(x, f_A(x))\ |\ x\in A\},\ \ \tilde{B} = \{(x, g_B(x))|\ x\in B\}$

1. $\tilde{A} \le \tilde{B} \leftrightarrow \boldsymbol{A \subseteq B}$
2. $f_A(x) \le g_B(x) \qquad \forall x \in A$

Definition 1.2.14: The standard Intersection and Union of two fuzzy sets:

Let $\tilde{A}, \tilde{B} \in I_4^X,$ *then*

$i)\ \tilde{A} \wedge \tilde{B} = \{(x\ , \min_{x\in A\cap B}(f_A(x), g_B(x)));\ \ \forall x \in A \cap B\}$

$$ii)\tilde{A} \vee \tilde{B} = \{(x, \max_{x \in A \cap B}(f_A(x), g_B(x)));\ \forall x \in A \cap B\}$$

Note 1.2.15: About the relation between 4th family and other families :

1. In case the sets which belong to 2^X are fixed this implies to the third family so the fourth family is considered more general than the third family .
2. In case the functions are fixed in other words all the functions is a one function and sets which belong to 2^X are vary , we get the first family so it is considered more general from the first family.
3. In case the sets which belongs to 2^X are fixed ,and the functions are vary we obtain the second family .

In other words the fourth family is the most general than the other families and the study of its properties and conclusion are more complicated.

From all above appears that it is necessary to define the fuzzy sets on all the points of the space in the first , second and fourth families respectively and it becomes as follows

$$f_A(x)=\begin{cases} 0 & if\ x \notin A \\ f(x) & if\ x \in A \end{cases}$$

Note 1.2.16 : As an extension for the first , second and fourth families ,we are introduced a new added families named fifth , sixth and seventh respectively as follows

Fifth family:

We introduce the fifth family , and denoted it by I_5^X and defined as follows:

$$I_5^X = \{\tilde{A}_i\ ;\ f: X \to I\ , \forall A \in 2^X\}$$

$$f_{A_i}(x)=\begin{cases} f_{A_i}(x) = 0 & \forall x \notin A_i \\ f_{A_i}(x) = f(x) & \forall x \in A_i \end{cases}$$

Example 1.2.17:

Return to example (1.2.2): we can rewrite Ã , $\tilde{B}$ as follows

$\tilde{A} = \{(1,1)\,, (2,0)\,, (3,0)\,, \dots, (100,0)\}$

$\tilde{B} = \{(1,1)\,,(2,\frac{1}{2})\,,(3,0)\,,(4,0)\,,\dots,\,(98,0)\,,(99,\frac{1}{99})\,,(100,0)\}$

Sixth Family:

we introduce the sixth family and denoted it by I_6^X and defined it as follows:

I_6^X={Ã ; $f_i \in \phi_A(x)$} , where

Ã={(x,$f_{iA}(x)$); $f_{iA}(x)$=0 , $\forall x \notin A$, $f_{iA}(x) = f_i(x)$, $\forall x \in A$}.

Example 1.2.18:

Return to example (1.2.6) in I_2^X: $\tilde{B}$, $\tilde{C}$ are rewritten as follows

$\tilde{B}$={(0,1) ,(1,$\frac{1}{2}$) ,(2,$\frac{1}{3}$) ,(3,0) ,(4,0) ,(5,0)}

$\tilde{C}$ ={(0,$\frac{1}{3}$) ,(1,$\frac{1}{5}$) ,(2,$\frac{1}{7}$) ,(3,0) ,(4,0) ,(5,0)}

Seventh Family:

We can introduce a seventh family and denoted it by I_7^X and defined it as follows

$$I_7^X=\left\{\tilde{A}_i\,, A_i = \left\{(x, f_{i_{A_i}}(x)), f_{i_{A_i}} \in \Psi_{A_i}(X)\,, f_{i_{A_i}}(x) = 0\ \forall x \notin A_i\,, f_{i_{A_i}} = f_i(x)\ \forall x \in A_i\right\}\right\}\}$$

It is considered the more general with respect to other families

Example 1.2.19 :

Return to example (1.2.12) , $\tilde{A}$ and $\tilde{B}$ are rewritten as follows :

$\tilde{A}$={$(x_1, 0)$, $(x_2, \frac{1}{8})$, $(x_3, 0)$, $(x_4, \frac{1}{24})$, $(x_5, 0)$, $(x_6, \frac{1}{48})$, $(x_7, 0)$, $(x_8, \frac{1}{80})$, $(x_9, 0)$}

$\tilde{B}$={$(x_1, 1)$, $(x_2, 0)$, $(x_3, \frac{1}{17})$, $(x_4, 0)$, $(x_5, \frac{1}{49})$, $(x_6, 0)$, $(x_7, \frac{1}{97})$) , $(x_8, 0)$, $(x_9, \frac{1}{161})$ }

We denote that $\tilde{X} = 1_x = \{(x, 1); \forall x \in X\}$ and $\tilde{\emptyset} = 0_x = \{(x, 0) \forall x \in X\}$ represent the absolute fuzzy set and the zero fuzzy set respectively

We appended the following illustrative examples about families:

Example 1.2.20 :

Let the universal of discourse X is the positive real numbers,0≤ x≤1

$\tilde{A}$={$(x, \mu_A(x))$ | $x \in X$}

$\tilde{B}$={$(x, V_B(x))$|$x \in X$}

where $\mu_A(x)$, $V_B(x)$ *are defined as* follows :

$$\mu_A(x) = \begin{cases} 0 & 0 \le x \le 0.5 \\ \dfrac{5(x-0.5)^2}{1+(x-0.5)^2} & 0.5 < x \le 1 \end{cases}$$

$$V_B(x) = \frac{1}{1+(x-0.7)^4} \quad 0 \le x \le 1$$

where $\tilde{A}$=damping ratio x considerably large than 0.5

$\tilde{B}$=damping ratio x approximately 0.7

In many branches of engineering the damping ratio is dimensionless measure describing two oscillations in a system after a disturbance .

$$\mu_A(x) \wedge V_B(x) = \begin{cases} \min\{(\frac{5(x-0.5)^2}{1+(x-0.5)^2} , \frac{1}{[1+(x-0.7)^4]}\} \\ 0 & x \le 0.5 \end{cases}$$

And $\mu_A(x) \vee V_B(x) = \max\{(\frac{5(x-0.5)^2}{1+(x-0.5)^2} , \frac{1}{[1+(x-0.7)^4]}\}$

Note that : An $A^c \neq \emptyset$, $A \cup A^c \neq X$

Where A^c represents the complements of A and also For the set B

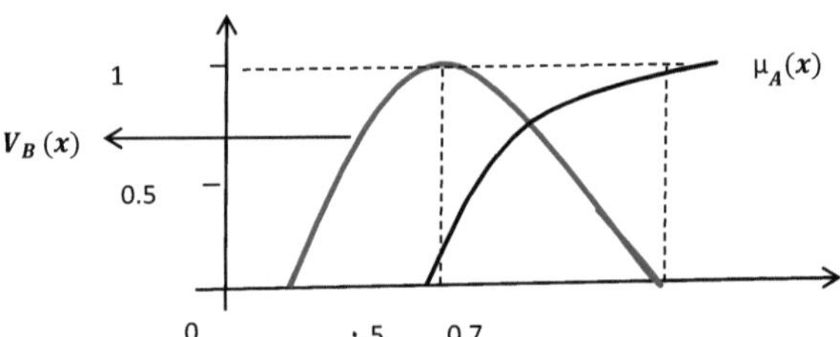

Figure (1-2) the graph of example 1.2.20

Example 1.2.21: The example is a modification for an example which is taken from , let

$$A_1(x) = \begin{cases} (x-2)+1 & x \in [1,2] \\ (2-x)+1 & x \in [2,3] \\ 0 & else\ where \end{cases}$$

$A_2(x)=\frac{1}{1+10(x-2)^2}$, $A_3(x)=e^{-|5(x-2)|}$, And

$$A_4(x)=\begin{cases} [1+\cos(2\pi(x-2))]/2 \\ \qquad when\ x \in [\frac{1}{2}, \frac{3}{2}] \\ 0 \qquad elsewhere \end{cases}$$

Represents the membership functions for fuzzy sets $\tilde{A}_1, \tilde{A}_2, \tilde{A}_3$ *and* $\tilde{A}_4$ 1 of real numbers close to 2.each of these fuzzy sets refer in particular ship for the general concept of a class of real numbers that are close to 2 as shown below :

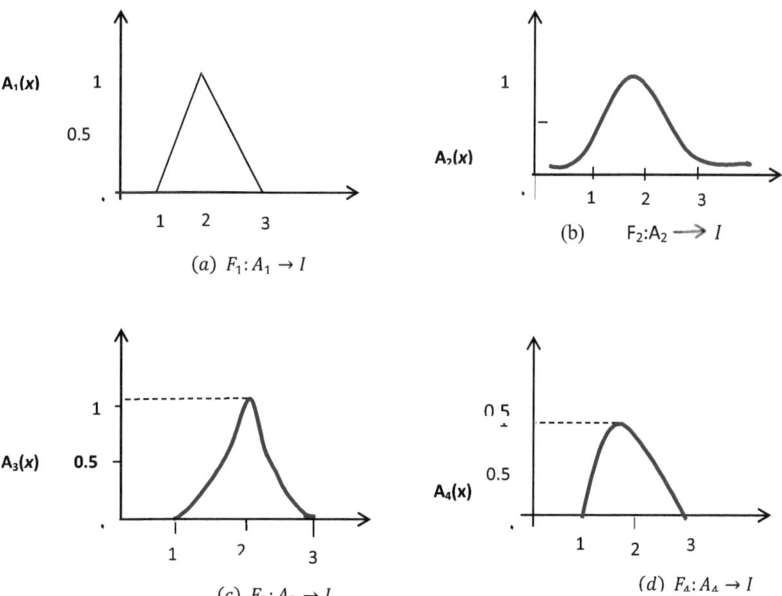

(a) $F_1: A_1 \rightarrow I$

(b) $F_2: A_2 \rightarrow I$

(c) $F_3: A_3 \rightarrow I$

(d) $F_4: A_4 \rightarrow I$

Figure (1-3) show the graphs of example 1.2.21

Example 1.2.22.

Let's assume that

$$\tilde{A} = \text{" x considerably} > 20\text{ "}$$
$$\tilde{B} = \text{" x approximately } 21\text{ "}$$

characterized by

$\tilde{A} = \{(x, f_A(x)) | x \in A\}$where

$$f_A(x) = \begin{cases} 0 & x \leq 20 \\ (1 + (x - 20)^{-2})^{-1} & x > 20 \end{cases}$$

and

$\tilde{B} = \{(x, f_B(x)) | x \in B\}$where

$f_{\tilde{B}}(x)$=$(1 + (x - 21)^4)^{-1}$,

$$f_{\tilde{A}\wedge\tilde{B}}(x) = \begin{cases} \min[(1 + (x - 20)^{-2})^{-1}, (1 + (x - 21)^4)^{-1}], & x > 20 \\ 0 & x \leq 20 \end{cases}$$

$$f_{\tilde{A}\vee\tilde{B}}(x) = \{\max[(1 + (x - 20)^{-2})^{-1}, (1 + (x - 21)^4)^{-1}], x \in X\}$$

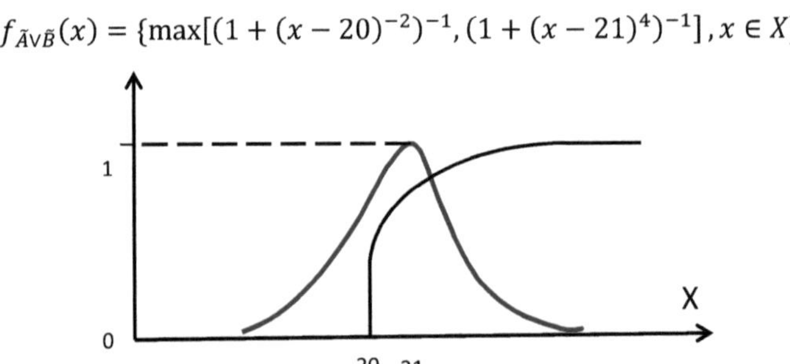

Figure (1-4) depict the example 1.2.22

1.3 Fuzzy topological space

In this section we introduce some definitions , Propositions and theorems concern the Fuzzy topology or it is called[0,1] topology which is an axiomatic subject.

Definition (1.3. 1)

Let X be a set. A fuzzy topology on X is a family ${}_fT$ of fuzzy sets in X, which satisfies the following conditions:

1) $\tilde{0}, \tilde{1} \in {}_fT$,
2) If $\tilde{A}, \tilde{B} \in {}_fT$, then If $\tilde{A} \wedge \tilde{B} \in {}_fT$
3) If $\tilde{A}_j \in {}_fT$ for each $j \in J$, then $\vee_{j \in J} \tilde{A}_j \in {}_fT$

${}_fT$ is called a fuzzy topology for X, and the pair $(\tilde{1}, {}_fT)$ is called a fuzzy topology space(in short ${}_fT$ fts) and $\tilde{1}$ is called fuzzy space. Every member of ${}_fT$ is called a fuzzy open set. A fuzzy set is called fuzzy closed iff its complement is fuzzy open.

Remark(1.3.2)
Let $(\tilde{1}, {}_f\tau)$be an fts . then :
1) $\tilde{0}, \tilde{1}$ are fuzzy closed sets.
2) If $\{\tilde{A}_j; j \in J\}$ is a family of fuzzy closed sets , then $\wedge_{j \in J} \tilde{A}_j$is a fuzzy closed set.
3) If $\tilde{A}, \tilde{B}$are fuzzy closed sets then $\tilde{A} \vee \tilde{B}$ is fuzzy closed set.

Now we introduce the definitions of base and sub base

Definition 1.3.3: A base for a fuzzy topological space $(\tilde{1}, {}_f\tau)$ is a sub collection β of ${}_f\tau$ such that member $\tilde{A} \in {}_f\tau$ can be written as $\tilde{A} = \vee_{j \in \Lambda} \tilde{A}_j$ where $\tilde{A}_j \in B$ and for some index Λ.

Definition 1.3.4: A sub base for a fuzzy topological space $(\tilde{1}, {}_f\tau)$ is a sub collection ζ of ${}_f\tau$ such that the collection of infimum of finite subfamilies of S forms a base for $(\tilde{1}, {}_f\tau)$.

Definition1.3.5: : A fuzzy set $\tilde{A}$ in a fts $(\tilde{1}, {}_f\tau)$ is called fuzzy nhd (or simply (f.nhd) of a fuzzy point p_x^λ if there is a $\tilde{B} \in \tau_f$ such that $p_x^\lambda \in \tilde{B} \leq \tilde{A}$. The family of all f.nhds of a fuzzy point p_x^λ is called f.nhd system of p_x^λ and denote by

$N_{p_x^\lambda}$ *alo* $A\ \tilde{}$ *is call a Mashhour nhd (denoted by* $M\_p\ (\tau)$*)*. Chang extended the concept of f.nhd fuzzy set as follows:

Definition 1.3.6 :

A fuzzy set $\tilde{A}$ in a fts $(\tilde{1}, {}_f\tau)$ is called f.nhd of a fuzzy set $\tilde{B}$ in $\tilde{1}$ if there is $\tilde{H} \in {}_f\tau$ such that $\tilde{B} \leq \tilde{H} \leq \tilde{A}$. The family consisting of all f.nhd of $\tilde{A}$ is called the f.nhd system of $\tilde{A}$.

Definition 1.3.7

A fuzzy set $\tilde{A}$ in fuzzy topological space $(\tilde{1}, {}_f\tau)$ is called quasi-nhd of P_λ^x $iff\ there\ exits\ \tilde{A}_1 \in \tau \ni \tilde{A}_1 \leq \tilde{A}$ and $P_x^\lambda q \tilde{A}_1$, the family of all fuzzy q-nbds of P_x^λ called the system of fuzzy q-nbds of P_x^λ and denoted by $N_{P_x^\lambda}^Q$.

Different nhd classes that have been introduced by several researcher like Ludescher , Kerre , Warren , Pu , Mashhour are stated in

The fuzzy singletons and fuzzy points reveal a complementary attitude towards the operations of Zadeh for the properties hold between the two points see .

Different nhd classes concepts that have been introduced by several researcher are stated below.

Definition1.3.8: Let $(\tilde{1}, {}_f\tau)$ be a fuzzy topological space, $\tilde{A}$ a fuzzy set in $\tilde{1}$, x an element of $\tilde{1}$,s a fuzzy singleton in $\tilde{1}$ and P_y^λ a fuzzy point is sub set of $X \times I$. then we redefine:

1. $\tilde{A}$ is Ludescher nhd of $x\,(denoted\ by\ \ L_x(\tau))\ iff\ (\exists \tilde{O} \in \tau)(x \in supp\ \tilde{O}\ and\ \tilde{O}\ \leq \tilde{A}\,)$.
2. $\tilde{A}$ is a Kerre nhd of $s\ (denoted\ by\ \mathcal{K}_x(\tau))\ iff\ \ (\exists \tilde{O}\ \in \tau)(s \subseteq \tilde{O} \leq \tilde{A}\,)$.
3. $\tilde{A}$ is a Warren nhd of x $(denoted\ by\ \ w_x(\tau))\ iff\ (\exists \tilde{O}\ \in \tau)(x \in supp\tilde{O}\ \ and\ \tilde{O}\ \leq \tilde{A}\,)$, and (f (x) = g(x)).
4. $\tilde{A}$ is a Pu nhd of $\tilde{s}$ $(denoted\ by\ P_s(\tau))\ iff\ \ (\exists \tilde{O}\ \in \tau)(s\ q\ \tilde{O}\ \ and\ \tilde{O} \leq \tilde{A}\,)$. [99]

Through our study we produce two new types of fuzzy points which are added to the type mentioned in [13] and via these two points we introduce a new fuzzy nhds denoted by ((L-A)) as (L-A) nhd below:

Definition 1.3.9

i- $\widetilde{A}$ is a (L-A) nhd of P_α $iff\ \exists\, \widetilde{O} \in {}_f\tau \ni P_\alpha \in \widetilde{O} \le \widetilde{A}$

ii- $\widetilde{A}$ is a (L-A) quasi nhd of P_α $iff\ \exists\, \widetilde{O} \in {}_f\tau \ni P_\alpha\, q\, \widetilde{O}\ and\ \widetilde{O} \le \widetilde{A}$

iii- $\widetilde{A}$ is a (L-A) nhd of P_α^B for $B \subseteq X$ $iff\ \exists\, \widetilde{O} \in {}_f\tau \ni P_\alpha^B \in \widetilde{O}\ and\ \widetilde{O} \le \widetilde{A}$

iv- $\widetilde{A}$ is a (L-A) quasi nhd of P_α^B for $B \subseteq X$ $iff\ \exists\, \widetilde{O} \in {}_f\tau \ni P_\alpha^B\, q\, \widetilde{O}\ and\ \widetilde{O} \le \widetilde{A}$

About the difference nhd classes and the relation between them, we noted that:

i. The nhd classes that satisfy the properties of classical nhd systems $C_1, C_2, C_3\ and\ C_4$ or not are illustrated as follows(to more details see [fuzzy of topology]):

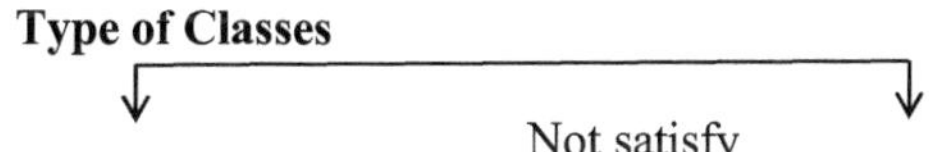

Satisfy C_1, C_2, C_3, C_4

1. $\mathcal{L}_x(\tau)$
2. $\mathcal{K}_x(\tau)$
3. $\mathrm{P}_s(\tau)$
4. $\mathrm{M}_p(\tau)$

Not satisfy

1. $w_x(\tau)$ only

ii. The characterization fuzzy topology by means of nhd systems for a classes of fuzzy sets (given as fuzzy sets nhd's) are as illustrated below:

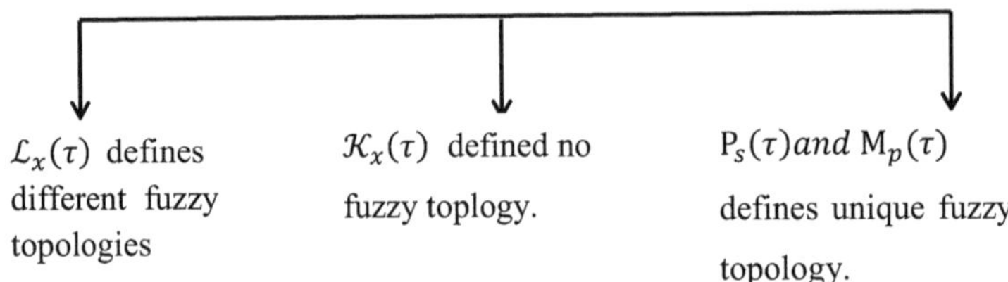

It may be concluded that only the approaches of Pu and Mashhor that totally induce a complete formal which is parallel with considered as the most proper extension of the classical description of a topology by means of nhd system

iii. The relations between $\mathrm{M}_p(\tau)$ of a fuzzy singleton , Pu nbd $\mathrm{P}_s(\tau)\ and\ \mathcal{K}_x(\tau)$ by using the complement fuzzy point concept are as follows :

$$\mathrm{M}_{x_\varepsilon}(\tau) = \mathrm{P}_{x_{1-\varepsilon}}(\tau)$$

$$\mathrm{M}_{x_\varepsilon}(\tau) = \cup_{\varepsilon'\in[\varepsilon,1]} \mathrm{k}_{\varepsilon'}(\tau)$$

Proposition 1.3.10: If $\tilde{A}$ is ((L-A)) nhd of the fuzzy point P_α iff $\tilde{A}$ is fnhd of fuzzy point $P_x^\alpha, \forall x \in X$.

Proof: There exists $\tilde{O} \in {}_fT, P_\alpha \tilde{\in} \tilde{O} \leq \tilde{A}$ iff $\alpha \leq f_o(x), \forall x \in X, \tilde{O} \leq \tilde{A}$ iff $P_x^\alpha \tilde{\in} \tilde{O}, \forall x \in X, \tilde{O} \leq \tilde{A}$ iff $\tilde{A}$ is a fuzzy nhd of $P_x^\alpha, \forall x \in X$.

Proposition1.3.11: If $\tilde{A}$ is ((L-A)) quasi nhd of fuzzy point P_α iff $\tilde{A}$ is a ((L-A)) quasi nhd of fuzzy point P_x^α for some $x \in X$.

Proof: Suppose that $\tilde{A}$ is ((L-A)) quasi nhd of fuzzy point P_α iff there exists a fuzzy open $\tilde{O} \in {}_fT, P_\alpha q \tilde{O}$ and $\tilde{O} \leq \tilde{A}$ iff $\alpha + f_o(x) > 1$ for some $x \in X$ and $\tilde{O} \leq \tilde{A}$ iff $P_\alpha q \tilde{O}$ for some $x \in X$ and $\tilde{O} \leq \tilde{A}$ iff $\tilde{A}$ is a fuzzy quasi nhd of a fuzzy point P_x^α for some $x \in X$.

Proposition 1.3.12: $\tilde{A}$ is fuzzy ((L-A)) nhd of a fuzzy point P_x^B iff $\tilde{A}$ is fuzzy ((L-A)) nhd of fuzzy point $P_x^B, \forall x \in B$.

Proof: Suppose that is fuzzy ((L-A)) nhd of a fuzzy point P_x^B iff there exists $\tilde{O} \in {}_fT, P_x^B \in \tilde{O}$ and $\tilde{O} \leq \tilde{A}$ iff $\alpha \leq f_o(x), \forall x \in B$ and $\tilde{O} \leq \tilde{A}$ iff $P_x^\alpha \in \tilde{O}, \forall x \in B$ and $\tilde{O} \leq \tilde{A}$ iff $\tilde{A}$ is fuzzy ((L-A)) nhd of fuzzy point $P_x^\alpha, \forall x \in B$.

Proposition1.3.13: A fuzzy set $\tilde{A}$ is fuzzy ((L-A)) quasi nhd of a fuzzy point P_x^B iff there exists $\tilde{A}$ is a quasi nhd of fuzzy point P_x^α for some $x \in B$.

Proof: Assume that $\tilde{A}$ is ((L-A)) quasi nhd of a fuzzy point $P_x^B, B \subset X$ iff there exists $\tilde{O} \in {}_fT, P_x^\alpha q\tilde{O}$ and $\tilde{O} \leq \tilde{A}$ iff $\alpha + f_o(x), x \in B$ and $\tilde{O} \leq \tilde{A}$ iff $P_x^B q\tilde{O}, \tilde{O} \leq \tilde{A}$ for some $x \in B$if $\tilde{A}$ is a qusi nhd of fuzzy point $P_x^\alpha$for some $x \in B$.

Proposition1.3.14:

Let $(\tilde{1}, {}_fT)$ be an fts. Then for each $p_x^\lambda \in FP(X), N_{p_x^\lambda}^Q$satisfies the followings:

1) p_x^λ is a q-coincident with A, for every $A \in N_{p_x^\lambda}^Q$.
2) If $\tilde{A}, \tilde{B} \in N_{p_x^\lambda}^Q$, then $\tilde{A} \wedge \tilde{B} \in N_{p_x^\lambda}^Q$.
3) If $\tilde{A} \in N_{p_x^\lambda}^Q$, and $\tilde{A} \leq \tilde{B}$ then $\tilde{B} \in N_{p_x^\lambda}^Q$.

Conversely, for each fuzzy point p_x^λ in X, if $N_{p_x^\lambda}^Q$ is the family of fuzzy sets in X satisfying the conditions (1),(2) and (3), then the family T of all fuzzy sets $\tilde{A}$ such that $\tilde{A} \in N_{p_x^\lambda}^Q$ whenever $p_x^\lambda q\tilde{A}$ is a fuzzy topology for X .

1.4 Fuzzy(limit point , adherence point, interior ,exterior and boundary) points

Definition1.4.1:

Let $\tilde{A}$ be a fuzzy set in an fts $(\tilde{1},\ _fT)$Then the fuzzy interior of $\tilde{A}$ is denoted $\tilde{A}^o$ and defined as the union of all fuzzy open subsets of X which are contained in $\tilde{A}$, i.e., $\tilde{A}^o = \vee\{\tilde{B} \in I^X : \tilde{B} \leq \tilde{A}, \tilde{B} \in\ _fT\}$.

Definition1.4.2:

Let $\tilde{A}$ be a fuzzy set in an fts $(\tilde{1},\ _fT)$Then the fuzzy closure of $\tilde{A}$ is denoted $\bar{\tilde{A}}$ and defined as the intersection of all fuzzy closed subsets of X which are containing in $\tilde{A}$, i.e., $\bar{\tilde{A}} = \wedge\{\tilde{A} \in I^X : \tilde{A} \leq \tilde{B}, \tilde{B}^c \in\ _fT\}$.

Proposition 1.4.3:

Let $\tilde{A}$ be a fuzzy set in an fts $(\tilde{1},\ _fT)$Then

a) $\tilde{A}^o$ is the largest fuzzy open set in X which is contained in $\tilde{A}$.
b) $\bar{\tilde{A}}$ is the smallest fuzzy closed set in X which is containing $\tilde{A}$.

Proposition1.4.4:

Let p_x^λ be a fuzzy point in X and $\tilde{A}$ be any fuzzy set in an fts $((\tilde{1},\ _fT))$, then $p_x^\lambda \in \bar{\tilde{A}}$ iff $\tilde{B}q\tilde{A}$ for every open set $\tilde{B}$ in X such that $\tilde{B}$ is q-coincident with p_x^λ.

Theorem1.4.5:

Let $\tilde{A}$ and $\tilde{B}$ be a fuzzy set in an fts $(\tilde{1},\ _fT)$. Then:

a) If $\tilde{B}$ is a fuzzy open sets, then $\tilde{B}q\tilde{A}$ iff $\tilde{B}q\bar{A}$.
b) $\tilde{A}$ is fuzzy open iff for each $P_x^r q\tilde{A}$, there exists $\tilde{B} \in N_{P_x^r}^Q$ such that $\tilde{B} \leq \tilde{A}$.

Proposition1.4.5:

For fuzzy sets $\tilde{A}$ and $\tilde{B}$ in an fts $\left((\tilde{1},\ _{f}T)\right)$, we have the following:

a) $\bar{0} = \tilde{0}; \bar{1} = \tilde{1}$.
b) $\tilde{A}$ is fuzzy open(resp. fuzzy closed) iff $\tilde{A}^{o} = \tilde{A}$ (resp. $\bar{\tilde{A}} = \tilde{A}$).
c) If $\tilde{A} \leq \tilde{B}$, then $\tilde{A}^{o} \leq \tilde{B}^{o}$(resp. $\bar{\tilde{A}} \leq \bar{\tilde{B}}$).
d) $\tilde{A}^{0^{0}} = \tilde{A}^{o}; \left(\bar{\bar{\tilde{A}}} = \bar{\tilde{A}}\right)$.
e) $\tilde{A}^{o}\wedge\tilde{B}^{o} = \left(\tilde{A}\wedge\tilde{B}\right)^{0}; \tilde{A}^{o}\vee\tilde{B}^{o} \leq \left(\tilde{A}\vee\tilde{B}\right)^{0}$.
f) $\overline{(\tilde{A}\wedge\tilde{B})} \leq \bar{\tilde{A}}\wedge\bar{\tilde{B}}; \overline{(\tilde{A}\vee\tilde{B})} = \bar{\tilde{A}}\vee\bar{\tilde{B}}$
g) $\tilde{A}^{c^{0}} = \left(\bar{\tilde{A}}\right)^{c}; \tilde{A}^{0^{c}} = \overline{\tilde{A}^{c}}$
h) $\bar{\tilde{A}}^{0} = \overline{\bar{\tilde{A}}^{0}}^{0}; \overline{\tilde{A}^{0}} = \overline{\overline{\tilde{A}^{0}}}^{0}$

Theorem1.4.6:

Let $\left((\tilde{1},\ _{f}T)\right)$ be an fts and let Y be a subset of X. Then the family $T_{Y} = \{\tilde{A}\wedge 1_{Y}: \tilde{A} \in T\}$ is a fuzzy topology on Y.

Definition 1.4.7 :
A fuzzy point $P_{x}^{\lambda}$is called an adherence point of a fuzzy set $\tilde{A}$ if and only if every q-nhd of P_{x}^{λ} is quasi coincident with $\tilde{A}$. The collection of all adherence points denoted by adh($\tilde{A}, P_{x}^{\lambda}$)

Definition 1.4.8:

A fuzzy point P_{x}^{α} is called:

1- Accumulation point of fuzzy set $\tilde{A}$ iff P_{x}^{α} is adherence point of $\tilde{A}$ and every Q-nhd of P_{x}^{α} is quasi-coincident and $\tilde{A}$ at some point different from $supp(P_{x}^{\alpha}) = x$, whenever $P_{x}^{\alpha} \tilde{\in} \tilde{A}$.The collection of all accumulation point of fuzzy set $\tilde{A}$ *denoted by* $ac(\tilde{A}, P_{x}^{\alpha})$
2- Interior point of fuzzy set $\tilde{A}$ iff there exists $\tilde{O} \in_{f}\tau$ such that$P_{x}^{\alpha} \in \tilde{O} \leq \tilde{A}$.

Definition 1.4.9: A fuzzy point $P_{\alpha}(P_{\alpha}^{B},\ B \subset X)$ is called

1- Adherence point of fuzzy set $\tilde{A}$ iff every (L-A) -Q-nhd of $P_\alpha(P_\alpha^B,$ $B \subset X)$ is quasi-coincident with $\tilde{A}$.

2- Adherence point P_x^α $(P_\alpha, P_\alpha^B,\ B \subset X)$ of fuzzy set $\tilde{A}$ iff every (L-A) –Q nhd of P_x^α $(P_\alpha, P_\alpha^B,\ B \subset X)$ is quasi-coincident with $\tilde{A}$.

3- d-Adherence point P_x^α $(P_\alpha, P_\alpha^B,\ B \subset X)$ of fuzzy set $\tilde{A}$ iff every (L-A) -nhd of $P_x^{1-\alpha}$ $(P_{1-\alpha}^B, P_{1-\alpha})$ is quasi-coincident with $\tilde{A}$.The collection of all d-adherence point of $\tilde{A}$ is denoted by $dad(\tilde{A}, P_x^{1-\propto})$

4- Interior point of fuzzy set $\tilde{A}$ iff there exists fuzzy open set $\tilde{O} \ni$ $P_\alpha(P_\alpha^B,\ B \subset X) \tilde{\in} \tilde{O}$, $\tilde{O} \leq \tilde{A}$.

5- Exterior point of fuzzy set $\tilde{A}$ iff its interior point of fuzzy set $\tilde{A}^c$.

6- A fuzzy point P_α is called

i- Accumulation point of fuzzy set $\tilde{A}$ iff every $(L$-A) nhd of $\tilde{N}$ of $P_\alpha \ni \alpha + f_N(x) > 1\ for\ some\ x \in B\ and\ f_N(y) + g_A(y) > 1\ for\ y \notin B$ .The collection of all accumulation points of $P_\propto$ $deoted\ by\ ac(\tilde{A}, P_\propto)$

ii- q- accumulation point of P_α of fuzzy set $\tilde{A}$ iff every (L-A) nhd of $\tilde{N}$ of $P_\alpha \ni f_n(y) + g_A(y) > 1$

$for\ y \notin B$ .The collection of all q-accumulation point of $P_\propto$ is denoted by $gac(\tilde{A}, P_\propto)$

iii- d- accumulation point of P_α^B of fuzzy set $\tilde{A}$ iff every (L-A) nhd of $\tilde{N}$ of $P_{1-\alpha}^B \ni$ $f_N(y) + g_A(y) > 1\ for\ y \notin B$.

7- Interior point of fuzzy set $\tilde{A}$ iff $\exists\, \tilde{O} \in {}_f\tau \ni P_\alpha(P_\alpha^B,\ B \subset X)$, $\tilde{\in}\ \tilde{O} \leq \tilde{A}$.

8- Exterior point of fuzzy set $\tilde{A}$ iff its interior of $\tilde{A}^c$.

9- The boundary point of the fuzzy set $\tilde{A}$ denoted by b($\tilde{A}$)= cl. $\tilde{A}$ - int. $\tilde{A}$

Proposition 1.4.10: A fuzzy point P_α is adherence point of fuzzy set $\tilde{A}$ iff for some $x \in X$, fuzzy point P_x^α is adherence point of fuzzy set $\tilde{A}$.

Proof: Assume that a fuzzy point P_α is adherence point of fuzzy set $\tilde{A}$ iff $\forall$ $\tilde{N} \in N^Q(P_\alpha) \ni P_\alpha q\tilde{N}$ and $\tilde{N}q\tilde{A}$ iff $\forall$ $\tilde{N} \in N^Q(P_\alpha), \alpha + f_N(x) > 1,$ $f_N(y) + g_A(y) > 1$ $for\ some\ x, y \in X$ iff for some $x \in X, P_x^\alpha q\tilde{N}$ and $\tilde{N}q\tilde{A}$ $\forall$ $\tilde{N} \in N^Q(P_x^\alpha)$ iff for some $x \in X$, fuzzy point P_x^α is adherence point of fuzzy set $\tilde{A}$.

Proposition 1.4.11: A fuzzy point $P_\alpha$is q-adherence point of fuzzy set $\tilde{A}$ iff $\forall\, x \in X$, a fuzzy point P_x^α is q-adherence point of fuzzy set $\tilde{A}$.

Proof: Assume that a fuzzy point P_α is q-adherence point of fuzzy set $\tilde{A}$ iff $\forall$ $\tilde{N} \in N(P_\alpha) \ni \tilde{N}q\tilde{A}$ iff $\forall$ (L – A)-nhd $\tilde{N}(P_\alpha), \alpha \le f_N(x)$, $\forall\, x \in X$ and $\tilde{N}q\tilde{A}$ iff $\forall\, x \in X$, $P_x^\alpha \in \tilde{N}$, $\tilde{N}q\tilde{A}$ $\in \forall$ (L – A)-nhd $\tilde{N} \in (L - A)N(P_x^\alpha)$ iff $\forall\, x \in X$, P_x^α is q-adherence point of fuzzy set $\tilde{A}$.

Proposition 1.4.12: A fuzzy point $P_\alpha$is d-adherence point of fuzzy set $\tilde{A}$ iff $\forall\, x \in X$, a fuzzy point P_x^α is d-adherence point of fuzzy set $\tilde{A}$.

Proof: Assume that a fuzzy point P_α is d-adherence point of fuzzy set $\tilde{A}$ iff $for\ each\ \tilde{N} \in (L - A)N(P_{1-\alpha}) \ni \tilde{N}q\tilde{A}$ iff $1 - \alpha \le f_N(x)$ $\forall\, x \in X \ni \tilde{N}q\tilde{A}$ $\forall\tilde{N} \in (L - A)N(P_{1-\alpha}^B)$ iff $\forall\, x \in X$ and $\tilde{N} \in (L - A)N(P_x^{1-\alpha}), \tilde{N}q\tilde{A}$ iff $\forall\, x \in X$, a fuzzy point P_x^α is d-adherence point of fuzzy set $\tilde{A}$.

Proposition 1.4.13: A fuzzy point P_x^B , $B \subset X$ is adherence point of fuzzy set $\tilde{A}$ iff for some $x \in B$, fuzzy point P_x^α is adherence point of fuzzy set $\tilde{A}$.

Proof: Assume that a fuzzy point P_α^B , $B \subset X$ is adherence point of fuzzy set $\tilde{A}$ iff for each $\tilde{N} \in (L - A)N^Q(P_\alpha^B)$ and $\tilde{N}q\tilde{A}$ iff $\alpha + f_N(x) > 1$, $f_N(y) + g_A(y) > 1$ $for\ some$ $x \in B$ and some $y \in X$ iff $for\ some$ $x \in B$ and $\forall$ $\tilde{N} \in (L - A)N^Q(P_x^\alpha)$ $\tilde{N}q\tilde{A}$ iff $for\ some$ $x \in B$, a fuzzy point P_x^α is adherence point of fuzzy set $\tilde{A}$.

Proposition 1.4.14: A fuzzy point P_x^B $(B \subset X)$ is q-adherence point of fuzzy set $\tilde{A}$ iff $\forall\, x \in B$, fuzzy point P_x^α is q-adherence point of fuzzy set $\tilde{A}$.

Proof: Assume that a fuzzy point P_x^B is q-adherence point of fuzzy set $\tilde{A}$ iff $\forall\, \tilde{N} \in (L-A)N(P_x^B)$ such that $\tilde{N}q\tilde{A}$ iff $\alpha \leq f_N(x), \forall\, x \in B\ and\ \tilde{N}q\tilde{A}$ iff $P_x^\alpha \in \tilde{N}, \forall \tilde{N} \in (L-A)N(P_x^B)\ and\ \tilde{N}q\tilde{A}, \forall\, x \in B$ iff $\forall\, x \in B, \tilde{N} \in (L-A)N(P_x^\alpha)\ and\ \tilde{N}q\tilde{A}$ iff $\forall\, x \in B$ a fuzzy point P_x^α is q-adherence point of fuzzy set $\tilde{A}$.

Proposition 1.4.15: A fuzzy point P_x^B is d-adherence point of fuzzy set $\tilde{A}$ iff $\forall\, x \in B$, fuzzy point P_x^α is d-adherence point of fuzzy set $\tilde{A}$.

Proof: Assume that a fuzzy point P_x^B is d-adherence point of fuzzy set $\tilde{A}$ iff $\forall\, \tilde{N} \in (L-A)N(P_{\alpha-1}^B)\ and\ \tilde{N}q\tilde{A}$ iff $\alpha - 1 \leq f_N(x), \forall\, x \in B\ and\ \tilde{N}q\tilde{A}$ iff $\forall\, x \in B\ \ P_x^{\alpha-1}\ \tilde{\in}\ \tilde{N}\ ,\ \tilde{N} \in (L-A)N(P_{1-\alpha}^B), \tilde{N}q\tilde{A}$ iff $\forall\, x \in B, \tilde{N} \in (L-A)N(P_x^{1-\alpha}) \ni \tilde{N}q\tilde{A}$ iff $\forall\, x \in B$, fuzzy point P_x^α is d-adherence point of fuzzy set $\tilde{A}$.

Proposition 1.4.16: If fuzzy point P_α^B is adherence point of fuzzy set $\tilde{A}$ then fuzzy point P_α is adherence point of fuzzy set $\tilde{A}$ for $B \subset X$.

Proof: Let a fuzzy point P_α^B is adherence point of fuzzy set $\tilde{A}$ iff $\forall\, \tilde{N} \in (L-A)N^Q(P_\alpha^B) \ni \tilde{N}q\tilde{A}$ iff

$\alpha + f_N(x) > 1 \quad for\ some\ \ x \in B \quad and\ \tilde{N}q\tilde{A}\ \ \forall\, \tilde{N} \in (L-A)N^Q(P_\alpha^B)$

But $B \subset X$, then $\alpha + f_N(x) > 1\, for\ some\ \ x \in X\ and\ \tilde{N}q\tilde{A}$, also $\tilde{N} \in (L-A)N^Q(P_\alpha^B)$ iff $\exists \tilde{O} \in_f \tau$ such that $P_\alpha^B q\tilde{O}$ and $\tilde{O}q\tilde{N}$ iff $\exists \tilde{O} \in_f \tau \ni \alpha + f_O(x) > 1$,
$f_O(y) + g_N(y) > 1\ for\ some\ \ x \in B$, but $B \subset X\ \ for\ some\ \ y \in X$, so $\exists \tilde{O} \in {}_f\tau \ni \alpha + f_O(x) > 1,\ f_O(y) + g_N(y) \quad > 1\ for\ some\ \ x, y \in X$

Therefore $\tilde{N}$ is Q-nhd of a fuzzy point P_α.

Hence the fuzzy point P_α is adherence point of fuzzy set $\tilde{A}$.

Proposition 1.4.17: If P_α is q-adherence point of fuzzy set $\tilde{A}$, then for $B \subset X$ fuzzy point P_α^B is q-adherence point of fuzzy set $\tilde{A}$.

Proof: Let a fuzzy point P_α is a q-adherence point of fuzzy set $\tilde{A}$ iff $\forall \tilde{N} \in (L-A)N(P_\alpha) \ni \tilde{N}q\tilde{A}$ iff $\forall \tilde{N} \in (L-A)N(P_\alpha), \exists \tilde{O} \in_f \tau \ni P_\alpha \in \tilde{O} \leq \tilde{N}$ and $\tilde{N}q\tilde{A}, \alpha \leq f_O(x) \leq g_N(x), \ \forall \ x \in X$, and $\tilde{A}q\tilde{N}$ then for each $B \subset X$, and $\forall \ x \in B$ iff $\alpha \leq f_O(x) \leq g_N(x)$, $\tilde{N}q\tilde{A}$, so $\forall \tilde{N} \in (L-A)N(P_\alpha), then \ \tilde{N} \in (L-A)N(P_\alpha^B)$, $\forall B \subset X$ and $\tilde{N}q\tilde{A}$. Therefore $\forall B \subset X$, fuzzy point P_α^B is q-adherence point of fuzzy set $\tilde{A}$.

Proposition 1.4.18: If P_α is d-adherence point of fuzzy set $\tilde{A}$, then for $B \subset X$ fuzzy point P_α^B is d-adherence point of fuzzy set $\tilde{A}$.

The proof of this proposition is similar technique of above proposition proof .

Proposition 1.4.19: A fuzzy point P_α is interior point of fuzzy set $\tilde{A}$ iff $\forall \ x \in X$, fuzzy point P_x^α **is** interior point of fuzzy set $\tilde{A}$.

Proof: Let a fuzzy point P_α is interior point of fuzzy set $\tilde{A}$ iff $\exists \tilde{O} \in_f \tau \ni P_\alpha \in \tilde{O} \leq \tilde{A}$ iff $\alpha \leq f_O(x) \leq g_A(x), \ \forall \ x \in X$ iff $\forall \ x \in X, P_x^\alpha \tilde{\in} \tilde{O} \leq \tilde{A}$ iff $\forall \ x \in X$, fuzzy point P_x^α is interior point of fuzzy set $\tilde{A}$.

Proposition 1.4.20: A fuzzy point P_α^B, $B \subset X$ is interior point of fuzzy set $\tilde{A}$ iff $\forall \ x \in B$, fuzzy point P_x^α is interior point of fuzzy set $\tilde{A}$.

Proof is a like the proof of above proposition.

Note that , a fuzzy point P_α is interior point of fuzzy set $\tilde{A}$ then P_α^B, $B \subset X$ **is** interior point of fuzzy set $\tilde{A}$.

Proposition 1.4.21: A fuzzy point P_α **is** exterior point of fuzzy set $\tilde{A}$ iff $\forall \ x \in X$, P_x^α **is** interior point of fuzzy set $\tilde{A}$.

Proposition 1.4.22: A fuzzy point P_α^B, $B \subset X$ is exterior point of fuzzy set $\tilde{A}$ iff $\forall \ x \in B$, fuzzy point P_x^α is exterior point of fuzzy set $\tilde{A}$.

The proof of above propositions is similarly technical proof of proposition 1.4.19.

We can use technical proof of propositions 1.4.11 to 1.4.18 to proof the following results .

Proposition 1.4.23: A fuzzy point P_α **is**

1- Accumulation point of fuzzy set $\tilde{A}$ iff for some $x \in X$, fuzzy point P_x^α is accumulation point of fuzzy set $\tilde{A}$.
2- q-accumulation point of fuzzy set $\tilde{A}$ iff $\forall\, x \in X$, fuzzy point P_x^α is q-accumulation point of fuzzy set $\tilde{A}$.
3- d-accumulation point of fuzzy set $\tilde{A}$ iff $\forall\, x \in X$, fuzzy point P_x^α is d-accumulation point of fuzzy set $\tilde{A}$.

Proposition 1.4.24: A fuzzy point $P_\alpha^B, B \subset X$ **is**

1- Accumulation point of fuzzy set $\tilde{A}$ iff for some $x \in B$, fuzzy point P_x^α is accumulation point of fuzzy set $\tilde{A}$.
2- q-accumulation point of fuzzy set $\tilde{A}$ iff $\forall\, x \in B$, fuzzy point P_x^α is q-accumulation point of fuzzy set $\tilde{A}$.
3- d-accumulation point of fuzzy set $\tilde{A}$ iff $\forall\, x \in B$, fuzzy point P_x^α is d-accumulation point of fuzzy set $\tilde{A}$.

Proposition 1.4.25: A fuzzy point $P_\alpha^B, B \subset X$ **is** accumulation point of fuzzy set $\tilde{A}$, then fuzzy point P_α **is** accumulation point of fuzzy set $\tilde{A}$.

Proposition 1.4.26: If a fuzzy point P_α is q-accumulation point of fuzzy set $\tilde{A}$, then for any , $B \subset X$, fuzzy point P_α^B is q-accumulation point of fuzzy set $\tilde{A}$.

Proposition 1.4.27: If a fuzzy point P_α is d-accumulation point of fuzzy set $\tilde{A}$, then for any , $B \subset X$, fuzzy point P_α^B is d-accumulation point of fuzzy set $\tilde{A}$.

1.5 Fuzzy separation axioms

There are many forms that researchers have defined the separation axioms of $T_\circ, T_1 and T_2$, but we will study carefully the definition of $T_\circ, T_1 and T_2$ and the same mechanism for the rest of the different definitions to the separation axioms.

Definition 1.5.1: A fuzzy topological space $(\tilde{1},_f\tau)$ is called

1- fuzzy $T_\circ - space$(or simply $f\tau_\circ - space$) w.r.t. fuzzy point P_x^α iff for any two different fuzzy point ,there exists fuzzy open set contains one of them.

2- fuzzy $T_1 - space$(or simply $f\tau_1 - space$) w.r.t. fuzzy point P_x^α iff for any different two fuzzy point ,there exists two fuzzy open sets contains one of these points not the other.

3- fuzzy $T_2 - space$(or simply $f\tau_2 - space$) w.r.t. fuzzy point P_x^α iff for any different two fuzzy point ,there exists two disjoint fuzzy open sets each of them contains one of these .

The above definition can be generalized with respect to point of kinds P_α and $P_\alpha^B, B \subset X$.

Definition 1.5.2: A fuzzy topological space $(\tilde{1},_f\tau)$ is called

1- fuzzy $T_\circ - space$ w.r.t. fuzzy point P_α (or $\mathrm{P}_\alpha^\mathrm{B}, \mathrm{B} \subset \mathrm{X}$) iff for any two different fuzzy point , there exists fuzzy open set contains one of them.

2- fuzzy $T_1 - space$ w.r.t. fuzzy point P_α (or P_x^α) iff for any different two fuzzy point ,there exists two fuzzy open sets contains one of these points not the other.

3- fuzzy $T_2 - space$ w.r.t. fuzzy point P_α(or P_x^α) iff for any different two fuzzy point ,there exists two disjoint fuzzy open sets each of them contains one of these points .

Also we can used the quasi-neighborhoods to define the separation axioms.

Definition 1.5.3: A fuzzy topological space $(\tilde{1}, {}_f\tau)$ is called

1- qf $T_\circ - space$ w.r.t. fuzzy point P_x^α (or P_α , P_α^B) iff for any two different fuzzy point , there exists Q-nhd of one of them and not contains the other.

2- qqf $T_\circ - space$ w.r.t. fuzzy point P_x^α (or P_α , P_α^B) iff for any two different fuzzy point , there exists Q-nhd of one of them and not quasi-conceited with other.

3- qf $T_1 - space$ w.r.t. fuzzy point P_x^α (or P_α , P_α^B) iff for any two different fuzzy point , there exists two Q-nhds of each of them, but each Q-nhds not contains the other points.

4- qqf $T_1 - space$ w.r.t. fuzzy point P_x^α (or P_α , P_α^B) iff for any two different fuzzy point , there exists two Q-nhds of each of these points , but these Q-nhds not quasi-conceited with other points.

5- qf $T_2 - space$ w.r.t. fuzzy point P_α or P_α^B iff for any two different fuzzy point , there exists two disjoint Q-nhds of these points .[2]

When we spoke about the difference between two fuzzy points of one type we noted that:

First: The difference two fuzzy points of kind P_x^α are

1. $P_x^\alpha \neq P_x^\lambda \ \ for \ \ \alpha \neq \lambda \in (0,1]$
2. $P_x^\alpha \neq P_y^\alpha \ \ for \ \ x \neq y \ \ in\, X$
3. $P_x^\alpha \neq P_y^\lambda \ \ for \ \ x \neq y \ \ in\, X \, and \ \ \alpha \neq \lambda \ \ in \ \ (0,1]$

Second : The different two fuzzy points of kind P_α

$P_\alpha \neq P_\lambda \;\; for \;\; \alpha \neq \lambda \;\; in\, (0,1)$

Third : The different two fuzzy points of kind $P_\alpha^B, B \subset X$ are

1- $P_\alpha^{\mathrm{B}} \neq P_\lambda^B \;\; for \;\; \alpha \neq \lambda \in (0,1]$
2- $P_\alpha^{\mathrm{B}} \neq P_\alpha^A \;\; for \;\; A \neq B \;\; in\, p(X)/\{\emptyset, X\}$
3- $P_\alpha^{\mathrm{B}} \neq P_\lambda^A \;\; for \;\; A \neq B \;\; in\, p(X)/\{\emptyset, X\}$ and $\alpha \neq \lambda \in (0,1]$.

Proposition 1.5.4 : $(\tilde{1}, {}_fT)$ is fT_i-space w.r.t fuzzy point $P_\alpha(P_\alpha^B)$ iff $(\tilde{1}, {}_fT)$ is fT_i-space w.r.t fuzzy point $P_x^\alpha, \forall x \in X(\forall x \in B), i = 0,1,2.$

Proof: Assume that $(\tilde{1}, {}_fT)$ is fT_o-space w.r.t fuzzy point P_α iff for each fuzzy points $P_\alpha \neq P_\lambda$, then $\alpha \neq \lambda, \forall x \in X, \exists \tilde{O} \in {}_fT$contains one of these fuzzy points not the other iff $P_x^\alpha \neq P_x^\lambda, \forall x \in X, \exists \tilde{O} \in {}_fT$ such that $P_x^\lambda \tilde{\in} \tilde{O}, P_x^\alpha \tilde{\notin} \tilde{O}$ or $P_x^\alpha \tilde{\in} \tilde{O}, P_x^\lambda \tilde{\notin} \tilde{O}$ iff $(\tilde{1}, {}_fT)$ is fT_o-space w.r.t fuzzy point $P_x^\alpha, \forall x \in X$. Similar can prove that for $i =$ 1or 2.

So from definitions 1.5.1 and 1.5.2 we get the following

Proposition 1.5.5:

1- If fuzzy topological space $(\tilde{1}, {}_f\tau)$ is qf $T_i - space$ w.r.t. fuzzy point of kind P_x^α, then the fuzzy topological space $(\tilde{1}, {}_f\tau)$ is f $T_i - space$ w.r.t. fuzzy point of kind P_α and $\mathrm{P}_\alpha^{\mathrm{B}}$ for all $B \subset X \; and \; i = 0,1,2$.
2- If fuzzy topological space $(\tilde{1}, {}_f\tau)$ is f $T_i - space$ w.r.t. fuzzy point of kind $\mathrm{P}_\alpha^{\mathrm{B}}$, then the fuzzy topological space $(\tilde{1}, {}_f\tau)$ is f $T_i - space$ w.r.t. fuzzy point of kind P_α .
3- If fuzzy topological space $(\tilde{1}, {}_f\tau)$ is qf $T_i - space$ w.r.t. fuzzy point of kind P_x^α or$\mathrm{P}_\alpha^{\mathrm{B}}$ then the fuzzy topological space $(\tilde{1}, {}_f\tau)$ is qf $T_i - space$ w.r.t. fuzzy point of kind P_α for any $B \subset X \; and \; i = 0,1,2$.
4- If fuzzy topological space $(\tilde{1}, {}_f\tau)$ is qqf $T_i - space$ w.r.t. fuzzy point of kind P_x^α or $\mathrm{P}_\alpha^{\mathrm{B}}$ then the fuzzy topological space $(\tilde{1}, {}_f\tau)$ is qqf $T_i - space$ w.r.t. fuzzy point of kind P_α for any $B \subset X \; and \; i = 0,1,2$.

Proof 3: For $i = 0$ let $P_\alpha \neq P_\lambda$ $then$ $\alpha \neq \lambda$ $for\ all\ x \in X$ so we get that$P_x^\alpha \neq P_x^\lambda$, but the fuzzy topological space $(\tilde{1}, {}_f\tau)$ is qf $T_\circ - space$,then $\exists\ \widetilde{N} \in N^Q(P_x^\alpha)\ and\ P_x^\lambda \notin \widetilde{N}$ that is $\alpha + f_N(x) > 1$ and $\lambda > f_N(x)$ therefore $P_\alpha q \widetilde{N}$ and $P_\lambda \notin \widetilde{N}$ this imply that $\forall\ \widetilde{N} \in N^Q(P_x^\alpha)$, then $\widetilde{N} \in N^Q(P_\alpha)$. Hence the fuzzy topological space $(\tilde{1}, {}_f\tau)$ is qf $T_\circ - space$ w.r.t. fuzzy point P_α.

Similarly the proof for $= 1,2$.

Proof 4: For $i = 0$ let $P_\alpha \neq P_\lambda$ $then$ $\alpha \neq \lambda$ $for\ all\ x \in X$ so we get that$P_x^\alpha \neq P_x^\lambda$, but the fuzzy topological space $(\tilde{1}, {}_f\tau)$ is qqf $T_\circ - space$, then $\exists\ \widetilde{N} \in N^Q(P_x^\alpha)\ and\ P_x^\lambda q\ \widetilde{N}$ that is $\alpha + f_N(x) > 1$ and $\lambda + f_N(x) \leq 1$. Then $P_\alpha\ q\ \widetilde{N}\ and\ P_\lambda\ q\ \widetilde{N}$ which imply that $\forall\ \widetilde{N} \in N^Q(P_x^\alpha)$ then $\widetilde{N} \in N^Q(P_\alpha)$ and $P_\lambda q\ \widetilde{N}$

So that the fuzzy topological space $(\tilde{1}, {}_f\tau)$ is qqf $T_\circ - space$ w.r.t. fuzzy point P_α. Similarly prove for $= 1,2$.

Proposition 1.5.6: A fuzzy topological space $(\tilde{1}, {}_f\tau)$ is f $T_\circ - space$ for each $P_x^\alpha \neq P_x^\lambda$ $(\alpha \neq \lambda\ and\ x \neq y)$, there exists fuzzy open set $\widetilde{U}$ such that $P_x^\alpha \widetilde{\in} \widetilde{U}$ and $P_x^\lambda\ q\ \widetilde{U}^c$.

Proof: Let $(\tilde{1}, {}_f\tau)$ is f $T_\circ - space$ iff $\forall\ P_x^\alpha \neq P_x^\lambda, \exists\ \widetilde{U} \in {}_f\tau \ni P_x^\alpha \widetilde{\in}\ \widetilde{U}$ and $P_x^\lambda \notin \widetilde{U}$ iff $\forall\ P_x^\alpha \neq P_x^\lambda$, $\exists\ \widetilde{U} \in {}_f\tau \ni P_x^\alpha \widetilde{\in}\ \widetilde{U}$ and $\lambda > f_u(y)$ iff $\forall\ P_x^\alpha \neq P_x^\lambda, \exists\ \widetilde{U} \in {}_f\tau \ni P_x^\alpha \widetilde{\in}\ \widetilde{U}$ and $\lambda + 1 - f_u(1) > 1$ iff $\forall\ P_x^\alpha \neq P_x^\lambda$, $\exists\ \widetilde{U} \in {}_f\tau \ni P_x^\alpha \widetilde{\in}\ \widetilde{U}$ and $P_x^\lambda\ q\ \widetilde{U}^c$.

Proposition 1.5.7: A fuzzy topological space $(\tilde{1}, {}_f\tau)$ is f $T_1 - space$ iff $\forall\ P_x^\alpha \neq P_y^\lambda$, $\exists\ \widetilde{U}, \widetilde{V} \in {}_f\tau \ni P_x^\alpha \widetilde{\in}\ \widetilde{U}, P_y^\lambda\ q\ \widetilde{U}^c, P_y^\lambda \in \widetilde{V}, P_x^\alpha\ q\ \widetilde{V}^c$.

Proposition 1.5.8: A fuzzy topological space $(\tilde{1}, {}_f\tau)$ is f $T_2 - space$ iff $\forall\ P_x^\alpha \neq P_y^\lambda, \exists\ \widetilde{U}, \widetilde{V} \in_f\tau, \ni P_x^\alpha \widetilde{\in}\ \widetilde{U}, P_y^\lambda \in \widetilde{V}, \widetilde{U} \wedge \widetilde{V} = \tilde{0},\ P_x^\alpha\ q\ \widetilde{V}^c, P_y^\lambda\ q\ \widetilde{U}^c$.

Definition 1.5.9: [2] A fuzzy topological space $(\tilde{1}, {}_f\tau)$ is called fuzzy compact iff every fuzzy open cover has finite sub cover .

The following diagram explain the relationship between the accumulation fuzzy points and adherence fuzzy points

Diagram(1-1)

The following diagrams explain the relationship between the interior

fuzzy points and exterior fuzzy points

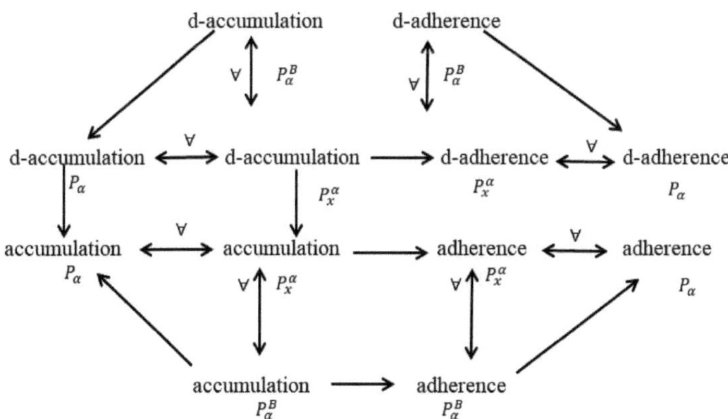

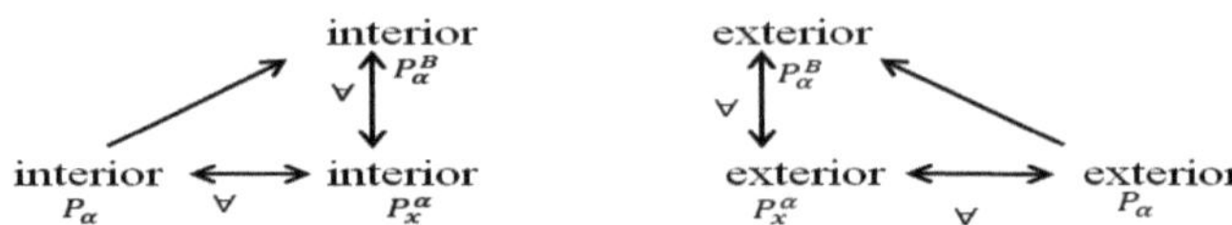

Diagram(1-2a) Diagram(1-2b)

The following diagrams explain the relationship between fuzzy separation axioms:

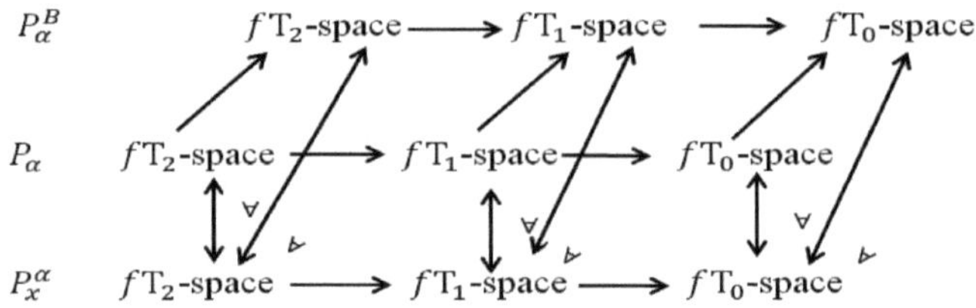

Diagram(1-3a)

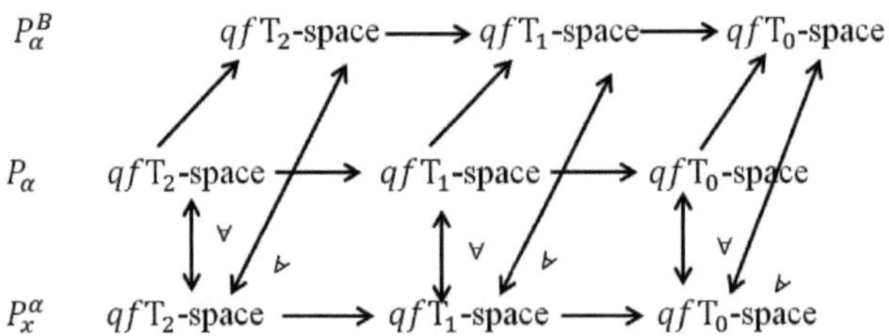

Diagram(1-3b)

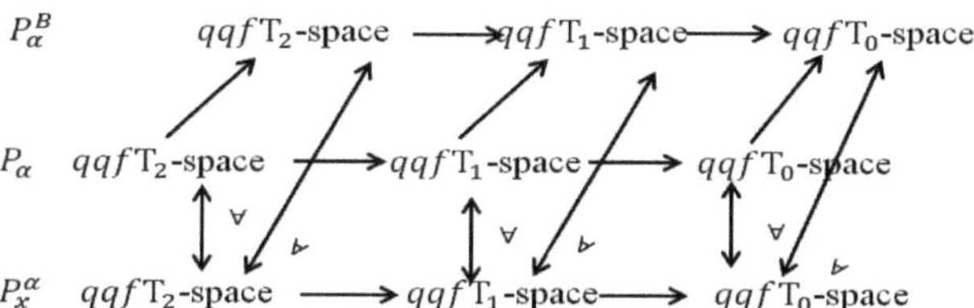

Diagram(1-3c)

Conclusions1.5.19:

1- In this study we have given definitions for three functions of fuzzy sets and obtain five new families of fuzzy sets which added to two families of [2,4,6,8] . After an analytical study for fuzzy sets , the families are classified into seven families which have an important role in solution of the uncertainty problems in real life to help the engineers and scientists. The seventh family is considered the more general family with respect to the other families . Also four new types of fuzzy nhds are introduced. We redefined the first type of fuzzy points and introduced two new types of fuzzy points with illustrative examples .Also we define several new types of quasi fuzzy (limit , adherent, interior , exterior and boundary)points and we define a new separation axioms and we study relationship between them.

2- Now we can study all theorems of (accumulation, adherence, interior, exterior and boundary) fuzzy points with respect to fuzzy point P_x^α of the types P_α and P_α^B.

3- It is easy to generlize all theorems of fuzzy separation exioms which have been studied on fuzzy point of type P_x^α to q-fuzzy separation exioms and qq-fuzzy separation exioms on fuzzy points of types P_α and P_α^B respectively.

1.6. Soft Sets

A new concept of science is appeared that is of soft set theory which was regarded as important tool and vivid concept to solve a plenty of problem in different domains as economics, engineering and communications and other branches of science. There are four different families of soft sets which are different features and properties but they are alike with other. We will give some definitions and the most important properties .

Definition 1.6.1:

A pair (F, A) denoted by F_A is called a soft set over X, where F is a mapping defined by $F: A \longrightarrow P(X)$, E is a set of all possible parameters with respect to X and $A \subseteq E$ such that $F_A = \{(e, F(e)), F(e) \in P(X), e \in A\}$.

Example 1.6.2:

Suppose that there are three houses in the universal set, $X = \{x_1, x_2, x_3\}$ under consideration and $A \subseteq E$, such that $A = \{(e_1 = \text{woody} \ , e_2 = \text{modren}\}$, y , consider the mapping F_A given by, $F_A(e_1)$ means "house(woody).

Let $F(e_1) = \{x_3\}, F(e_2) = \{x_1, x_2\}$, so

$F_A = \{ (e_1, \{x_3\}), (e_2, \{x_1, x_2\})\}$.

Remark 1.6.3:

Let $A = \{e_1, e_2, \dots , e_n\}$ be a set of parameters. Then negation set of A is denoted by $\neg A$

Proposition 1.6.4:

1. $\neg(\neg A) = A$.
2. $\neg(A \cup B) = \neg A \cup \neg B$.
3. $\neg(A \cap B) = \neg A \cap \neg B$.

Families of a Soft Sets 1.6.5:

Four families which are mentioned in [34] with an illustrative examples for each family is presented in this subsection .

First Family:

Let X be any universal set and let E be a set of all parameters of X. Then we will denote for this family as $SS_1(X)$ such that:
$SS_1(X) = \{F_{iA_i}, where\ A_i \subseteq E\ and\ F_i: A_i \longrightarrow p(X)\}$.

Example1.6.6:

Suppose that there are five phones in the universal set X by which operate given by $X = \{x_1, x_2, x_3, x_4, x_5\}$ and let E be set of parameters which operated by many plat forms , such that $E_X = \{e_1 = Amocos, e_2 = windos\ MS, e_3 = Soloris, e_4 = QNX, \quad e_5 = Macintosh\}$.
And let $A, B, \subseteq E_X$ such that $A = \{e_1, e_3\}$, $B = \{e_2, e_4\}$,
Then we can define soft sets F_A, G_B over the common universal X such that $F: A \longrightarrow P(X)$, $F_A = \{(e_1, \{x_1\}), (e_3, \{x_1, x_2\})\}$, and $G: B \longrightarrow P(X)$, where $G_B = \{(e_2, \{x_1, x_5\}), (e_4, \{x_3\})\}$.
Note that the soft sets F_A and G_B belong to $SS_1(X)$.

Definition 1.6.7:

The soft union of any two soft sets F_A and G_B is a soft set H_C where C=A$\cup B$ and $\forall c \in C$

$$H_C(c) = \begin{cases} F_A(c) & c \in A - B \\ G_B(c) & c \in B - A \\ F_A(c) \cup G_B(c) & c \in A \cap B \end{cases}$$

and we write the union as below:

$$H_C(c) = F_A(c) \tilde{\cup} G_B(c), \forall c \in C$$

Definition 1.6.8:

The soft intersection of any two soft sets F_A and G_B is a soft set H_C where

$$H_C(c) = F_A(c) \tilde{\cap} G_B(c), \forall c \in C = A \cap B$$

and write the intersection as below: $F_A \ \tilde{\cap}\, G_B = H_C$

Definition 1.6.9:

For a soft set F_A over the universal set X the relative complement of F_A denoted by $(F_A)^c = {F_A}^c$ where $F^c: A \to P(X)$ is a mapping given by $F^c(a) = X - F(a)$, $\forall a \in A$

Example1.6.10: Let the universal set $U = \{u_1, u_2, u_3\} and\ A = \{e_1, e_2, e_3\}$, $B = \{e_2, e_3\}, C = \{e_1, e_3\}$. Let the soft sets$F_A$ $=\{(e_1, U), (e_2, U), (e_3, U)\}$, $G_B = \{(e_2, \{x_1, x_2\}), (e_3, \{x_1\})\}$. And $H_C = \{(e_1, \{\emptyset\}), (e_3, \{\emptyset\})\}$, then

$F_A \,\tilde{\cap}\, G_B = \{(e_2, \{x_1, x_2\}), (e_3, \{x_1\})\}$

$F_A \,\tilde{\cup}\, G_B = \{(e_1, U), (e_2, U), (e_3, U)\}$

${F_A}^c = \{(e_1, \{\emptyset\}), (e_2, \{\emptyset\}), (e_3, \{\emptyset\})\}$.

Proposition 1.6.11:

Let F_A be a soft set over the universal set X. Then

1. $F_A \,\tilde{\cap}\, \tilde{\varphi} = \tilde{\varphi}$.
2. $F_A \,\tilde{\cup}\, \tilde{\varphi} = F_A$.
3. $F_A \,\tilde{\cap}\, \tilde{X} = F_A$.
4. $F_A \,\tilde{\cup}\, \tilde{X} = \tilde{X}$.
5. $F_A \,\tilde{\cup}\, F_A = F_A$
6. $F_A \,\tilde{\cap}\, F_A = F_A$

Second Family:

This family is denoted by $SS_2(X)$ at the formula as follows: $SS_2(X) = \{F_{A_i}, where\ A_i \subseteq E\ and\ F: A_i \longrightarrow P(X)\}$.

Example 1.6.12:

Let E be the set of all possible parameters that determine student levels and let X be the universal set consisting of three students such that $X = \{x_1, x_2, x_3\}$ and A, B and C are sub sets of E as followes:

$A = \{a_1{:}\, good, a_2{:}\, pass\}$,$B = \{a_2{:}\, pass, a_3{:}\, Excellent\}$ and $C = \{a_1{:}\, good, a_3{:}\, Excellent\}$.

Then the soft sets:

$F_A = \{(a_1, \{x_1\}), (a_2, \{x_3\})\}$,$F_B = \{(a_2, \{x_3\}), (a_3, \{x_2\})\}$ and $F_C = \{(a_1, \{x_1\}), (a_3, \{x_2\})\}$ belongs to $SS_2(X)$.

Third Family: (Central Soft Set)

We will denote it by $SS_3(X)$ and defines as follows:

$SS_3(X) = \{F_{/A_i}, where\ A_i \subseteq E\ and\ F{:}\, E \longrightarrow P(X)\}$.

Where $F_{/A_i}(e){:}\, F(e)$ for all $e \in A_i$.

Definition 1.6.13:

Let A be a subset of the parameters set E. A pair (F, A) is called a soft set with a central set A, where F is a mapping given by $F{:}\, A \longrightarrow P(X)$.

For simply, we call $(F_{/A}, A)$ a central soft set. For two central soft sets $(F_{/A}, A)$ and $(G_{/B}, B)$, we say $(F_{/A}, A) = (G_{/B}, B)$, if $F_{/A} = G_{/B}$ and $A = B$.

Example 1.6.14:

Let $X = \{x_1, x_2, x_3, x_4, x_5\}$ be a different models of laptops. Let $E = \{e_1, e_2, e_3, e_4\}$ represents the price of a laptops . Consider two central soft sets $(F_{/A}, A)$ and $(G_{/B}, B)$, such that: $F_{/A}{:}\, E_X \longrightarrow P(X)$ and $G_{/B}{:}\, E_X \longrightarrow P(X)$ defined over X, where $A = \{e_2, e_3\}$ and $B = \{e_1, e_4\}$, so $F_{/A}$ and $G_{/B}$ are two central soft sets .The mappings $F_{/A}$ and $G_{/B}$ are defined as follows:

$F(e_1) = \{x_2\}, F(e_2) = \{x_1, x_5\}, F(e_3) = \{x_1\}, F(e_4) = \{x_3, x_4, x_5\},$

$G(e_1) = \{x_1, x_2\},\ G(e_2) = \{x_2, x_3, x_4, x_5\},\ G(e_3) = \{x_4\},\ G(e_4) = \{x_5\}$ and. So $F_{/A}(e_2) = \{x_1, x_5\}, G_{/B}(e_1) = \{x_1, x_2\}, G_{/B}(e_4) = \{x_5\}$, and $F_{/A}(e_3) = \{x_1\}$.

Fourth Family:

The study will be on this family and we denote by $SS_4(X)$, which is as follows:

$SS_4(X) = \{F_{i_A}, where\ A \subseteq E\ and\ F_i: A \longrightarrow P(X)\}$

Example 1.6.15:

Let $X = \{x_1, x_2, x_3\}$ and $A \subseteq E$, $A = \{e_1, e_2\}$ then F_A and G_A are soft sets over X where $F_A = \{(e_1, \{x_2\}), (e_2, \{x_1, x_3\})\}$ and $G_A = \{(e_1, \{x_2, x_3\}), (e_2, X)\}$.

Definition 1.6.16:

A soft set F_A over X is said to be:

1. Null soft set if $F(e) = \varphi$, for all $e \in E$, and denoted by $\tilde{\varphi}$. that means $\tilde{\varphi} = \{(e, \varphi): e \in E\}$.
2. Absolute soft set if $F(e) = X$, for all $e \in E$, and denoted by $\tilde{X}$, that means $\tilde{X} = \{(e, X): e \in E\}$.

Definition 1.6.17:

There are three types of the soft points in $\tilde{X}$ as follows:

1. The first type of the soft point in $\tilde{X}$ denoted by F_e^x, where

   $$F_e^x(a) = \begin{cases} \{x\} & \text{if } a = e \\ \varphi & \text{if } a \neq e \end{cases}.$$

2. The second type of the soft point in $\tilde{X}$ denoted by F_x, where

$F_x = \{(e, \{x\}) : e \in A\}$.

3. The third type of the soft point in $\tilde{X}$ denoted by F_e, where

$$F_e(a) = \begin{cases} F(e) \neq \varphi & \text{if } a = e \\ \varphi & \text{if } a \neq e \end{cases} \quad \text{for all } e \in A.$$

Definition 1.6.18:

A soft point F_e^x is said to be belonging to the soft set G_A, denoted by $F_e^x \tilde{\in} G_A$, if $F(e) \subseteq G(e)$.

Proposition 1.6.19:

The union of any collection of soft points can be considered as a soft set and every soft set can be expressed as union of all soft points belonging to it. (i.e. $F_A = \bigcup_{F_e^x \tilde{\in} F_A} F_e^x$).

Definition 1.6.20: Two soft sets $G_{A,} H_A$ in $SS_1(X)$ are called disjoint $G_A \tilde{\cap} H_A = \emptyset_A$ if $G_{e,} \cap H_e = \emptyset$ for all $e \in E$.

Definition 1.6.21: Two soft points F_{e_1} and F_{e_2} are in X are distinct if their corresponding soft sets G_A and H_A are disjoint.

1.7 Soft Topological Spaces

In this section soft topology is studded with some properties as closure , interior , limit points, base and sub base and soft separation axioms are introduced

Definition 1.7.1:

Let X be an initial universal set and $A \subseteq E$. Let ${}_s\tau$ be the collection of soft sets over X, then ${}_s\tau$ is said to be a soft topology on X if:

1. $\tilde{\varphi}, \tilde{X}$ belong to ${}_s\tau$.
2. The union of any number of soft sets in ${}_s\tau$ belongs to ${}_s\tau$.
3. The intersection of any two soft sets in ${}_s\tau$ belongs to ${}_s\tau$.

The triplet $(\tilde{X},\ _s\tau, A)$ is called a soft topological space over X. The members of $_s\tau$ are called soft open sets and a soft set F_A is called soft closed iff the soft complement $(\tilde{X} \tilde{\setminus} F_A)$ is soft open.

Example 1.7.2:

Let $X = R$ be the real numbers, let $A = R^+$ be the positive real numbers.

Then $_s\tau = \{F_{An}: for\ all\ e \in A, for\ all\ n \in \mathbb{N}\ where:$ $F_{An} = \{(e, \{e + n: n \in \mathbb{N}\})\} \cup \{\tilde{\varphi}, \tilde{X}\}\}$.

Hence $(\tilde{X},\ _s\tau, A)$ is a soft topological space.

Definition 1.7.3: Let $(X,\ _s\tau)$ be a soft topological space. **A** sub collection β of $_s\tau$ is called a base for a soft topological space $(X,\ _s\tau)$ if every member of $_s\tau$can be expressed as union of members of β.

Definition 1.7.4: Let $(X,\ _s\tau)$ be a soft topological space. A sub collection S of $_s\tau$is said to be a sub base of $_s\tau$ if the family of all finite intersections of members of S forms a base for $(X,\ _s\tau)$.

Definition 1.7.5:

Let $(\tilde{X},\ _s\tau, A)$ be a soft topological space and let Y be a non-empty subset of X. Then $\tilde{\tau}_Y = \{F_A \tilde{\cap} \tilde{Y}: F_A \tilde{\in}\ _s\tau\}$ is said to be the soft relative topology on Y and $(\tilde{Y},\ _s\tau_Y, A)$ is called soft sub space of $(\tilde{X},\ _s\tau, A)$.

Definition 1.7.6:

Let $(\tilde{X},\ _s\tau, A)$ be a soft topological space and let $F_A \tilde{\in} SS_1(X)$ and $F_e^x \tilde{\in} \tilde{X}$. Then:

1. F_A is called a soft neighborhood (briefly S- nhd) of F_e^x, if there exists $G_A \tilde{\in}\ _s\tau$ such that $F_e^x \tilde{\in} G_A \tilde{\subseteq} F_A$.

2. F_A is called a soft open nhd of F_e^x, if $F_A \tilde{\in} {}_s\tau$ and F_A is S- nhd of F_e^x.

The family of all S- nhd of F_e^x is denoted by $N_{F_e^x}$. And the family of all soft open nhd of F_e^x is denoted by $N_{{}_s\tau(F_e^x)}$ and similarly we can define the neighborhood w.r.t the others soft points.

Example 1.7.7:

Let $= \{x_1, x_2, x_3\}$, $A \subseteq E$ where $A = \{e_1, e_2\}$ and define $F_{1A} = \{(e_1, \{x_2\}), (e_2, \{x_3\})\}$, $F_{2A} = \{(e_1, \{x_1, x_2\}), (e_2, \{x_3\})\}$, $F_{3A} = \{(e_1, \varphi), (e_2, \{x_3\})\}$. Let ${}_s\tau = \{\tilde{\varphi}, \tilde{X}, F_{1A}, F_{2A}, F_{3A}\}$ be a soft topological space over X and let $G_A = \{(e_1, \{x_1, x_2\}), (e_2, \{x_2, x_3\})\}$

we note $F_{2A} \tilde{\subseteq} G_A$ and since $x_3 \in F_2(e_2)$ hence $F_{e_2}^{x_3} \tilde{\in} F_{2A}$ so $F_{e_2}^{x_3} \tilde{\in} F_{2A} \tilde{\subseteq} G_A$ thus G_A is a S-nhd of the soft point $F_{e_2}^{x_3}$. But $G_A \tilde{\notin} {}_s\mathcal{T}$ so G_A is not soft open nhd of $F_{e_2}^{x_3}$.

Theorem 1.7.8:

Let $(\tilde{X}, {}_s\mathcal{T}, A)$ be a soft topological space and $N_{F_e^x}$ is a nhd system of F_e^x . Then

1. If $F_A \tilde{\in} N_{F_e^x}$, then $F_e^x \tilde{\in} F_A$.
2. If $G_A \tilde{\in} N_{F_e^x}$, and $G_A \tilde{\subseteq} F_A$, then $F_A \tilde{\in} N_{F_e^x}$.
3. If $F_A, G_A \tilde{\in} N_{F_e^x}$ then $F_A \tilde{\cap} G_A \tilde{\in} N_{F_e^x}$.

Definition 1.7.9:

Let $(\tilde{X}, {}_s\mathcal{T}, A)$ be a soft topological space, $F_A \tilde{\in} SS_1(X)$ and $F_e^x \tilde{\subseteq} \tilde{X}$. If every open nhd of F_e^x intersect F_A in some soft points other than F_e^x itself, then F_e^x is called a soft limit point of F_A.

The set of all limit points of F_A which is called derived sets of F_A is denoted by $d(F_A)$.

In other word: $F_e^x \tilde{\in} d(F_A)$ iff $F_A \tilde{\cap} (G_A \tilde{\setminus} F_e^x) \neq \tilde{\varphi}$ for all $G_A \tilde{\in} {}_s\mathcal{T}(F_e^x)$.

Theorem 1.7.10:

Let $(\tilde{X}, \widetilde{{}_s\mathcal{T}}, A)$ be a soft topological space, $F_A, G_A \tilde{\in} SS_1(\mathcal{X})$. Then

1. $d(F_A) \tilde{\subseteq} c\ell(F_A)$.
2. $F_A \tilde{\subseteq} G_A$, then $d(F_A) \tilde{\subseteq} d(G_A)$.
3. $d(F_A \tilde{\cap} G_A) \tilde{\subseteq} d(F_A) \tilde{\cap} d(G_A)$
4. $d(F_A \tilde{\cup} G_A) = d(F_A) \tilde{\cup} d(G_A)$.
5. $F_A$is a soft closed iff $d(F_A) \tilde{\subseteq} F_A$.

Definition 1.7.11:

Let $(\tilde{X}, {}_s\mathcal{T}, A)$ be a soft topological space. And $F_A \tilde{\in} SS_1(X)$. Then

1. The Soft closure of the soft set F_A is denoted by $c\ell(F_A)$ is the intersection of all soft closed super set of F_A. $c\ell(F_A)$ is the smallest soft closed set over X which contains F_A.

(i.e. $c\ell(F_A) = \tilde{\cap}\{H_A: H_A$ is closed soft set and $F_A \tilde{\subseteq} H_A\}$.

2. The Soft interior of the soft set F_A is denoted by $int(F_A)$ is defined as the union of all soft open sets contained in F_A. Thus $int(F_A)$ is the largest soft open set contained in F_A.

(i.e. $int(F_A) = \tilde{\cup}\{H_A: H_A$ is a soft open set and $H_A \tilde{\subseteq} F_A\}$.

Definition 1.7.12:

Let $(\tilde{X}, {}_s\mathcal{T}, A)$ be a soft topological space and let F_A be a soft set over X. Then

1. F_A is soft closed iff $c\ell(F_A) = F_A$.
2. F_A is soft open iff $int(F_A) = F_A$.

Theorem 1.7.13:

Let $(\tilde{X}, {}_{s}\mathcal{T}, A)$ be a soft topological space and let F_A, G_A be soft sets over X. Then

1. $c\ell(\tilde{\varphi}) = \tilde{\varphi}, c\ell(\tilde{X}) = \tilde{X}$.
2. $F_A \tilde{\subseteq} c\ell(F_A)$.
3. If F_A is closed, then $c\ell(c\ell(F_A)) = F_A$.
4. $F_A \tilde{\subseteq} G_A$ implies $c\ell(F_A) \tilde{\subseteq} c\ell(G_A)$.
5. $c\ell(F_A \tilde{\cup} G_A) = c\ell(F_A) \tilde{\cup} c\ell(G_A)$.
6. $c\ell(F_A \tilde{\cap} G_A) \tilde{\subseteq} c\ell(F_A) \tilde{\cap} c\ell(G_A)$.

Theorem 1.7.14 :

Let $(\tilde{X}, {}_{s}\mathcal{T}, A)$ be a soft topological space and let F_A, G_A be soft sets over X. Then

1. $int(\tilde{\varphi}) = \tilde{\varphi}. int(\tilde{X}) = \tilde{X}$.
2. $int(F_A) \tilde{\subseteq} F_A$.
3. $int(int(F_A)) = int(F_A)$.
4. $F_A \tilde{\subseteq} G_A$ impiies $int(F_A) \tilde{\subseteq} int(G_A)$.
5. $int(F_A \tilde{\cap} G_A) = int(F_A) \tilde{\cap} int(G_A)$.
6. $int(F_A) \tilde{\cup} int(G_A) \tilde{\subseteq} int(F_A \tilde{\cup} G_A)$.

Theorem 1.7.15:

Let $(\tilde{X}, {}_{s}\mathcal{T}, A)$ be a soft topological space. A soft point $F_e^x \tilde{\in} c\ell(F_A)$ iff each S- nhd of F_e^x intersects F_A.

Note 1.7.16:

Let $(\tilde{X}, \tilde{\mathcal{T}}, A)$ be a soft topological space, and $F_A \tilde{\in} SS_1(X)$ then $F_e^x \tilde{\in} c\ell(F_A \tilde{\setminus} F_e^x)$ iff $F_e^x \tilde{\in} d\big(d(F_A) \tilde{\setminus} F_e^x\big)$.

Proof:

$F_e^x \tilde{\in}\ c\ell(F_A \tilde{\setminus} F_e^x) = \big(d(F_A) \tilde{\setminus} F_e^x\big) \tilde{\cup}\, d\big(d(F_A) \tilde{\setminus} F_e^x\big)$iff

$F_e^x \tilde{\in} d\big(d(F_A) \tilde{\setminus} F_e^x\big)$.

Now we introduce the new following Soft separation axioms

Definition 1.7.17: A soft topological space $(\tilde{X},\ {}_s\tau)$ is called

1. Soft $T_\circ - space$(or simply ${}_s\tau_o - space$) w.r.t. soft point F_e iff for any two different soft point ,there exists soft open set contains one of them.
2- Soft $T_1 - space$(or simply ${}_s\tau_1 - space$) w.r.t. soft point F_e iff for any different two soft point ,there exists two soft open sets contains one of these points not the another.
3- Soft $T_2 - space$(or simply ${}_s\tau_2 - space$) w.r.t. soft point F_e iff for any distinct two soft point ,there exists two disjoint soft open sets each of them contains one of them .

Noted that the definition above can be generalized with respect to points of type F_x and F_e^x.

Definition 1.7.18: A soft topological space $(\tilde{1}, {}_s\,\tau)$ is called

1- Soft $T_\circ - space$ w.r.t. soft point F_x or F_e^x iff for any two different soft point, there exists soft open set contains one of them.
2- Soft $T_1 - space$w.r.t. soft point F_x or F_e^x iff for any different two soft point, there exists two soft open sets contains one of these points not the another.
3- Soft $T_2 - space$ w.r.t. soft point F_x or F_e^x iff for any different two soft point, there exists two disjoint soft open sets each of them contains one of these points .

Noted that about the distinct of two soft points of the same type, we illustrate the following:

i. For soft point of type F_e^x

1. $F_e^x \neq F_e^y \; for \; x, y \in X$
2. $F_{e_1}^x \neq F_{e_2}^x \; for \; e_1, e_2 \in A$
3. $F_{e_1}^x \neq F_{e_2}^y$

ii. For soft point of type F_x

$F_x \neq F_y$

iii. For soft point of type F_e

$F_{e_1} \neq F_{e_2}$

Proposition 1.7.19:

1. Every soft ${}_s\tau_1$- space is a soft ${}_s\tau_o$-space.
   Proof: Straightforward
2. Every soft τ_2-space is soft τ_1-space
   Proof: If $(\tilde{X}, {}_s\tau)$ is a soft τ_2-space then by definition for $F_e^x, F_e^y \tilde{\in} X_A, F_e^x \neq F_e^y$, there exists soft open sets F_A, G_A such that $F_e^x \in F_A$ and $F_e^y \in G_A$ and $F_A \tilde{\cap} G_A = \emptyset$. Since $F_A \tilde{\cap} G_A = \emptyset$. $F_e^y \tilde{\notin} F_A$ and $F_e^x \tilde{\notin} G_A$. Thus $(\tilde{X}, {}_s\tau)$ is ${}_s\tau_1$-space.

Definition 1.7.20 :

A soft topological space is called soft compact iff every soft cover has finite sub cover .

Theorem 1.7.21 :

A soft topological space is soft compact iff every family of soft closed with the property of finite intersection property has non-empty intersection .

Theorem 1.7.22 :

Every soft closed of soft compact topological spaces is soft compact .

Exercises :

1- Find the pseudo inverse and fuzzy complement of the following increasing generators

a- g(a) = a

b- g(a) = $\frac{1}{\gamma}\ ln(1+\gamma a)$

c- g(a) = $\frac{1}{\gamma}\ln(1+\gamma a^{w})$

d- g(a) = $\frac{a}{\gamma+(1-\gamma)a}$ $\qquad \gamma > 0$

2- Find the pseudo inverse and fuzzy complement of the following decreasing generator

a- h(x) = m (1- x) $\qquad$ m > 0

b- h(x) = $1 - x^{v}$ $\qquad$ v > 0

3- find the t-norm and t-conorm of the following increasing generators

a- f(x) = $1-(1-x)^{q}$ $\quad q \neq 0$

b- f(x) = $-\ln(\frac{m^{1-x}-1}{m-1})$ $\qquad m > 0\ ,\ m \neq 1$

4- Let X = { x_1 , x_2 , x_3 , x_4 , x_5 } ,

$\tilde{A}$ = { (x_1,0) , (x_2 ,0.1) , (x_3 , 0.4) , (x_4 , 0) , (x_5 , 0.1)}

$\tilde{B}$ ={ (x_1, 0.7), (x_2 , 0) , (x_3 , 0.6) , (x_4 , 0) , (x_5 , o.1)}

$\tilde{C}$ ={ (x_1 , 0) , (x_2 , 0) , (x_3 , 0.4) , (x_4 , 0) , (x_5 , 0.1) }

$\tilde{D}$ = { (x_1 , 0.7) , (x_2 , 0.1) , (x_3 , 0.6) , (x_4 , 0) , (x_5 , o.1) }

${}_{f}\tau = \{\tilde{0}, \tilde{1}, \tilde{A}, \tilde{B}, \tilde{C}, \tilde{D}\}$, E = { x_2 , x_3 } , β = 0.013 and ∝ = 0.12 .

Find the following

a- $N\ (P^{\propto})$

b- $q - N(P^{\propto})$

c- L–N(P_{β}^{E}) $and\ L - N(P_{\propto})$

d- $L - Nq(P_{\beta}^{E})$ $\ and\ L - Nq(P_{\alpha})$

e- $L_{x_5}({}_{f}\tilde{\tau})$ $\quad and \quad L_{x_2}({}_{f}\tilde{\tau})$

f- *Let* $\tilde{F}$ = { (x_1 , 0.2) , (x_2 , 0) , (x_3 , 0.8), (x_4 , 0) , (x_5 , 0.01) }

Find $K_x({}_{f}\tilde{\tau})$ for fuzzy set $\tilde{F}$.

5- Let X = { 1 , 2 , 3 , 4 , 5 , 6 , 7 } , A= { 1 , 2 , 3 } , B = { 3 , 4 , 5 , 6 } , C = { 3 } , D = { 1, 2 , 3 , 4 , 5 , 6 }

$$f_A(x) = \frac{x}{x^2+1} \quad , h_C(x) = g_B(x) = e^{-x} \ ,$$

$$k_D(x) = \begin{cases} f_A(x) & if\ x \in \{1,2,3\} \\ g_B(x) & if\ x \in \{4,5,6\} \end{cases}$$

And ${}_f\tilde{\tau} = \{ \tilde{0}, \tilde{1}, \tilde{A}, \tilde{B}, \tilde{C}, \tilde{D} \}$ .Find int ($\tilde{S}$) *and cl* ($\tilde{S}$)

$\tilde{S} = \{ (1, 0.01), (2, 0), (3, 0.1), (4, 0.5), (5, 0.4), (6, 0.6), (7,0) \}$

6- Show that , for any fuzzy point $P_\propto \in cl(\tilde{A})$ iff $\forall \tilde{B} \in {}_f\tilde{\tau}, \tilde{B} q \tilde{A}$ *and* $P_\alpha q \tilde{B}$.

7- For any B subset of X , $0 < \propto < 1$, and $P_\propto^B \leq cl(\tilde{A})$, show that for any fuzzy open set $\tilde{D}$ *we have* $P_\propto^B q \tilde{D}$.

8- Prove that the propositions 1.4.24, 25 , 26 and 1.4.27

9- Prove that in fuzzy fT_2 topological space

$P_x^\propto = \wedge \{ \tilde{O} \in {}_f\tilde{\tau} : P_x^\alpha \in \tilde{D} \}$.

10-prove Theorem 1.7.10 part 3 and give example explain that the converse may be not true .

11-Prove the Theorem 1.7.14 part 6 and give example explain that the converse may be not true .

12- prove that the soft subspace of sT_i – space is sT_i – space for i=0 , 1 , 2 .

13- prove that the soft sT_0 – space iff the soft closure of any two different soft points are different .

14- if we have soft Huasdorff space , show that the intersection of all soft closed sets containing an arbitrary soft point is equal to this soft point .

Chapter Two

Fuzzy Soft Sets and Fuzzy Soft Topological Spaces

2. Preliminaries:

In this chapter some definitions in view of the mathematical analysis concepts about the composition between the soft sets and fuzzy sets are presented , a new different types of fuzzy soft points produced via the composition between the three different types of the fuzzy points and the different three types of soft points with illustrative examples. In view of the mathematical analysis concepts this impose that the definitions must be becomes more precise and simple. For the composition between the soft sets and fuzzy sets to find a fuzzy soft set such that its membership function is a composition of the fuzzy function with the soft function , say f is a function from $P(X)$ to $I = [0,1]$ and F is a function from E to P(X) where X is a universal set , E is the set of parameters of the universal set X . The families that contains these sets which are represent the composition of soft families with fuzzy families to introduce a fuzzy families which depends on them are classified . [36] mentioned four families of soft sets denoted by $SS_1(X), SS_2(X), SS_3(X)\ and\ SS_4(X)$ with some basic notions . Three different soft points mentioned in [34] which are related with the first soft family $SS_1(X)$. We redefined a type of fuzzy point which was produced by [13], and we produced two new other types of fuzzy points.

A new different fuzzy soft points appear which are produced via the composition between the three different types of the fuzzy points which produced by [26] with the different three types of soft points mentioned in [36] and we get a new fuzzy soft points with illustrative examples.

Via the composition between the fuzzy topology and soft topology a fuzzy soft topological spaces and base and sub base of fuzzy soft spaces is obtained and introduced a new fuzzy soft separation axiom with theorems related with the fuzzy soft points further a definition of fuzzy soft compact space with theorem are presented.

2.1 The Fuzzy Sets and Fuzzy Points with Respect to $P(X)$

In this section it is worth to mention that we are presented some new definitions which are concern our study .

Definition 2.1.1 :
Let X be a universe set and $I = [0,1]$. The function $f: p(X) \to I$ is called the membership of a fuzzy set $\tilde{A}$.

Definition 2.1.2 :

Let X be a universe set and $P(X)$ be the power set of X.

Let's denote that

1. $M(P(X), I) = \{f; f: P(X) \to I\}$ be the set of all functions f on the power of the universe set X to the unit interval $I = [0,1]$.

2. $\Phi_{\mathcal{L}}(P(X)) = \{f_{\mathcal{L}}: \forall f \in M(P(X), I)\}, f_{\mathcal{L}}(B) = f(B), \forall B \in \mathcal{L}$ be the set of all restriction functions in $M(P(X), I)$ with respect to the subset $\mathcal{L}$ of $P(X)$.

3. $\psi_{\mathcal{L}}(P(X)) = \{f_{i\mathcal{L}}: \forall f_i \in \Phi_{\mathcal{L}}(P(X))\}, \forall \mathcal{L} \subseteq P(X)$ where $f_{i\mathcal{L}}(B) = f_i(B), \forall B \in \mathcal{L}_i$ be the set of all restriction functions in $M(P(X), I)$ with respect to the sub collections $\mathcal{L}_i$ of $P(X)$.

So the family of all fuzzy sets with respect to $P(X)$ and $I = [0,1]$is:

$I_7^{P(X)} = \left\{\tilde{A}_i: \tilde{A}_i = \left(B, f_{i\mathcal{L}_i}(B)\right), f_{\mathcal{L}_i} \in \psi_{\mathcal{L}_i}(P(X)), i \in \Lambda\right\}$ for any index Λ

Definition 2.1.3:

For any fuzzy sets $\tilde{A}, \tilde{B} \in I_7^{P(X)}$ we create the fuzzy intersection, union and complement with respect to $P(X)$,

$$\tilde{A} \cap \tilde{B} = \left\{\left(C, min\left(f(C), g(C)\right)\right): C \in P(X)\right\},$$

$$\tilde{A} \cup \tilde{B} = \left\{\left(C, max\left(f(C), g(C)\right)\right): C \in P(X)\right\},$$

$$\tilde{A}^c = \{(C, 1 - f(C)): \forall C \in P(X)\}$$

$$\tilde{0} = \{(B, 0): \forall B \in P(X)\},$$

$$\tilde{1} = \{(B, 1): \forall B \in P(X)\}.$$

Types of fuzzy points 2.1.4 :

Type I: For some $A \subset P(X)$ and $0 < \alpha \leq 1$, we say that P_α^A is classical fuzzy point on the fuzzy point of type I if

$$P_\alpha^A(B) = \begin{cases} \alpha, if\ B = A \\ 0, if\ B \neq A \end{cases}$$

For any $A, B \in P(X)$

Type II: Let $\emptyset \notin \mathcal{L} \subset P(X)$ and for some $0 < \alpha < 1$. The formula $P_\alpha^{\mathcal{L}}$ is called the fuzzy point of Type II where

$$P_\alpha^{\mathcal{L}}(B) = \begin{cases} P_\alpha^B, & if\ B \in \mathcal{L} \\ 0, & if\ B \notin \mathcal{L} \end{cases}$$

Type ɪɪɪ : For any $0 < \alpha < 1$, the formula P_α is called the fuzzy point of type ɪɪɪ such that $P_\alpha(\mathrm{A}) = \alpha\ \forall\ \mathrm{A} \in P(X)$.

Note that $P_\alpha = \bigcup_{A \in \mathcal{L}} P_\alpha^A$.

2.2 The Composition of Fuzzy Sets with Soft Sets

We introduce the following definition to explain the composition of fuzzy sets with soft sets with examples.

Definition 2.2.1:

Let X be an initial universe set , E be a set of parameters of elements of the set X and $I = [0,1]$. The fuzzy soft sets (simply ${}_f F_A$) with membership $\#F$, where $\#F: E \to I$ such that $\#F(e) = (f \circ F)(e), \forall e \in E, f: P(X) \to I$ and $F: E \to P(X)$. So ${}_f\tilde{A}_S = \{(e, F(e)); e \in E\}$.

Example 2.2.2:

Let X=$\{x_1, x_2, x_3\}$, E =$\{e_1, e_2\}$

$F(e_1) = \{x_1, x_2\}$, $F(e_2) = \{x_3\}$

$f(\{x_1\}) = 0.1$, $f(\{x_2\}) = 0.2$, $f(\{x_3\}) = 0.14$

$f(\{x_1, x_2\}) = 0.11$, $f(\{x_1, x_3\}) = 0.12$

$f(\{x_2, x_3\}) = 0.13$, $f(X) = 1$, $f(\phi) = 0$

Then $(f \circ F)(e_1) = f(F(e_1)) = f(\{x_1, x_2\}) = 0.11$

$(f \circ F)(e_2) = f(F(e_2)) = f(\{x_3\}) = 0.14$

Thus ${}_f\tilde{A}_s = \{(e_1, 0.11), (e_2, 0.14)\}$

To find the fuzzy soft point with respect to the function $\#F: E \to I$ which depends on the composite of the memberships of fuzzy points in 1.1.14 and the types of soft points 1.6.17 , we get the following:

I) 1-The composite of fuzzy point P_α^A with the soft point F_e^x and if $A = \{x\}$ we have the fuzzy soft point P_α^e

2- If $x \notin A$, then the composite is the null fuzzy soft set.

Example 2.2.3 :

Let

$X = \{x_1, x_2\}, E = \{e_1, e_2\}, \quad P_{0.2}^{A} = P_{0.2}^{\{x_2\}} = \{(\{x_1\}, 0), (\{x_2\}, 0.2), (\emptyset, 0), (X, 0)\},$

$P_{0.2}^{X} = \{(\{x_1\}, 0), (\{x_2\}, 0), (\emptyset, 0), (X, 0.2)\}, F_{e_1}^{x_2} = \{(e_1, \{x_2\}), (e_2, 0)\}.$

Then the composite of $P_{0.2}^{\{x_2\}}$ with $F_{e_1}^{x_2}$, i.e. $\{x_2\} = A$.

$(f \circ F)(e_1) = f(F(e_1)) = f(\{x_2\}) = 0.2, (f \circ F)(e_2) = f(F(e_2)) = f(\emptyset) = 0.$

That is, $\{(e_1, 0.2), (e_2, 0)\} = P_{0.2}^{e_1}$.

The composite of $P_{0.2}^{X}$ with $F_{e_1}^{x_2}$,

$(f \circ F)(e_1) = f(F(e_1)) = f(\{x_2\}) = 0 \;, (f \circ F)(e_2) = f(F(e_2)) = f(\emptyset) = 0.$

$\{(e_1, 0), (e_2, 0)\} = \tilde{0}$ the null fuzzy soft set.

II) The composite of fuzzy point P_{α}^{A} with soft point F^{x} is

$$P_{\alpha}^{A} \circ F^{x} = \begin{cases} P_{\alpha}, & if\ \{x\} = A \\ \tilde{0}, & if\ x \notin A \end{cases}$$

Example 2.2.4:

Let X= $\{x_1, x_2\}, E = \{e_1, e_2\}$,

$$P_{0.2}^{\{x_1\}} = \{(\{x_1\}, 0.2), (\{x_2\}, 0), (\emptyset, 0), (X, 0)\},$$

$$F^{x_1} = \{(e_1, \{x_1\}), (e_2, \{x_1\})\},$$

$$F^{x_2} = \{(e_1, \{x_2\}), (e_2, \{x_2\})\}$$

$P_{0.2}^{\{x_1\}} \circ F^{x_1} = \{(e_1, 0.2), (e_2, 0.2)\} = P_{0.2}$ because

$f(F(e_1)) = f(\{x_1\}) = 0.2$ and $f(F(e_2)) = f(\{x_1\}) = 0.2$, $P_{0.2}^{\{x_1\}} \circ F^{x_2} = \tilde{0}$ because $f(F(e_1)) = f(\{x_2\}) = 0$ and $f(F(e_2)) = f(\{x_2\}) = 0$

III) The composite of fuzzy point P_α^A with soft point F_e is

$$P_\alpha^A \circ F_e = \begin{cases} P_\alpha^e, & if\ A = F(e) \\ \tilde{0}, & if\ A \neq F(e) \end{cases}$$

Example 2.2.5:

Let $X = \{x_1, x_2\}, E = \{e_1, e_2\}$,

$P_{0.2}^{\{x_2\}} = \{(\{x_1\}, 0), (\{x_2\}, 0.2), (\emptyset, 0), (X, 0)\}$,

$$F_{e_1} = \{(e_1, X), (e_2, \emptyset)\}, F_{e_2} = \{(e_1, \emptyset), (e_2, \{x_2\})\},$$

$P_{0.2}^{\{x_2\}} \circ F_{e_1} = \tilde{0}$ because $f(F(e_1)) = f(X) = 0, and\ f(F(e_2)) = f(\emptyset) = 0$

$$P_{0.2}^{\{x_2\}} \circ F_{e_2} = \{(e_1, 0), (e_2, 0.2)\} = P_{0.2}^{e_2},$$

$$f(F(e_1)) = f(\emptyset) = 0, and\ f(F(e_2)) = f(\{x_2\}) = 0.2$$

IV) The composite of fuzzy point P_α^C with soft point F_e^x where $C \subset P(X)$.

1. If $\emptyset \in C$ and $\{x\} \in C$, then $P_\alpha^C \circ F_e^x = P_\alpha$
2. If $\emptyset \in C$ and $\{x\} \notin C$, then $P_\alpha^C \circ F_e^x = P_\alpha^B, for\ e \notin B \subset E$
3. If $\emptyset \notin C$ and $\{x\} \in C$, then $P_\alpha^C \circ F_e^x = P_\alpha^e$
4. If $\emptyset \notin C$ and $\{x\} \notin C$, then $P_\alpha^C \circ F_e^x = \tilde{0}$

Example 2.2.6:

Let $X = \{x_1, x_2\}, E = \{e_1, e_2, e_3\}, C_1 = \{\{x_1\}, \emptyset\}, C_2 = \{\{x_2\}, X\}$,

$P_{0.2}^{C_1} = \{(\{x_1\}, 0.2), (\{x_2\}, 0), (\emptyset, 0.2), (X, 0)\}$,

$P_{0.2}^{C_2} = \{(\{x_1\}, 0), (\{x_2\}, 0.2), (\emptyset, 0), (X, 0.2)\}$.

$F_{e_1}^{x_1} = \{(e_1, \{x_1\}), (e_2, \emptyset), (e_3, \emptyset)$

$F_{e_1}^{x_2} = \{(e_1, \{x_1\}), (e_2, \emptyset), (e_3, \emptyset)\}$

$P_{0.2}^{C_1} \circ F_{e_1}^{x_1} = \{(e_1, 0.2), (e_2, 0.2), (e_3, 0.2) = P_{0.2}$ because

$f(F(e_1)) = f(\{x_1\}) = 0.2$

$f(F(e_2)) = f(\emptyset) = 0.2$ and $f(F(e_3)) = f(\emptyset) = 0.2$

$P_{0.2}^{C_2} \circ F_{e_1}^{x_2} = \{(e_1, 0.2), (e_2, 0), (e_3, 0) = P_{0.2}^{e_1}$, because

$f(F(e_1)) = f(\{x_2\}) = 0.2$

$f(F(e_2)) = f(\emptyset) = 0$and $f(F(e_3)) = f(\emptyset) = 0$

Also $P_{0.2}^{C_1} \circ F_{e_1}^{x_2} = \{(e_1, 0), (e_2, 0.2), (e_3, 0.2) = P_{\propto}^{B}$, $B = \{e_2, e_3\}$, because

$f(F(e_1)) = f(\{x_2\}) = 0$

$f(F(e_2)) = f(\emptyset) = 0.2$ and $f(F(e_3)) = f(\emptyset) = 0.2$

$P_{0.2}^{C_2} \circ F_{e_1}^{x_1} = \{(e_1, 0), (e_2, 0), (e_3, 0) = 0, becouse$

$f(F(e_1)) = f(\{x_1\}) = 0$

$f(F(e_2)) = f(\emptyset) = 0$ and $f(F(e_3)) = f(\emptyset) = 0$

V) The composite of fuzzy point P_α^C with soft point F^x , where $C \subset P(X)$ we have that

$$P_\alpha^C \circ F^x = \begin{cases} P_\alpha & \text{if } \{x\} \in C \\ \tilde{0} & \text{if} \{x\} \notin C \end{cases}$$

Example 2.2.7:

Let $F^{x_1} = \{(e_1, \{x_1\}), (e_2, \{x_1\}), (e_3, \{x_1\})$ and $P_{0.2}^{C_1}, P_{0.2}^{C_2}$ from example 2.2.6 we have

$P_{0.2}^{C_1} \circ F^{x_1} = \{(e_1, 0.2), (e_2, 0.2), (e_3, 0.2)\} = P_{0.2}$, because

$f\big(F(e_1)\big) = f(\{x_1\}) = 0.2$, $f\big(F(e_2)\big) = f\big(F(e_3)\big) = f(\{x_1\}) = 0.2$

$P_{0.2}^{C_2} \circ F^{x_1} = \{(e_1, 0), (e_2, 0), (e_3, 0)\} = \tilde{0}$ because $f\big(F(e_1)\big) = f(\{x_1\}) = 0$

and $f\big(F(e_2)\big) = f\big(F(e_3)\big) = f(\{x_1\}) = 0$

VI) The composite of fuzzy point P_α^C with soft point F_e , when $C \subset P(x)$ we have that:

1. If $\emptyset \in C$ and $F(e) \in C$, then $P_\alpha^C \circ F_e = P_\alpha$
2. If $\emptyset \in C$ and $F(e) \notin C$, then $P_\alpha^C \circ F_e = P_\alpha^B, e \notin B \subset E$
3. If $\emptyset \notin C$ and $F(e) \in C$, then $P_\alpha^C \circ F_e = P_\alpha^e$
4. If $\emptyset \notin C$ and $F(e) \notin C$, then $P_\alpha^C \circ F_e = \tilde{0}$

Example 2.2.8:

Let $F_{e_1} = \{(e_1, X), (e_2, \emptyset), (e_3, \emptyset)\}$

$F_{e_2} = \{(e_1, \emptyset), (e_2, \{x_2\}), (e_3, \emptyset)\}$ and take $P_{0.2}^{C_1}$ and $P_{0.2}^{C_2}$ from example 2.2.6 , so we have

$P_{0.2}^{C_1} \circ F_{e_1} = \{(e_1, 0), (e_2, 0.2), (e_3, 0.2)\} = P_{0.2}^B, B = \{e_2, e_3\}$ because

$f(F(e_1)) = f(X) = 0$ and $f(F(e_2)) = f(F(e_3)) = f(\emptyset) = 0.2$

$P_{0.2}^{C_2} \circ F_{e_1} = \{(e_1, 0), (e_2, 0), (e_3, 0)\} = \tilde{0}$, because

$f(F(e_1)) = f(X) = 0$ and $f(F(e_2)) = f(F(e_3)) = f(\emptyset) = 0$

$P_{0.2}^{C_1} \circ F_{e_2} = \{(e_1, 0.2), (e_2, 0.2), (e_3, 0.2)\} = P_{0.2}$, because

$f(F(e_1)) = f(\emptyset) = 0.2$, $f(F(e_2)) = f(\{x_2\}) = 0.2$, and

$f(F(e_3)) = f(\emptyset) = 0.2$

$P_{0.2}^{C_2} \circ F_{e_2} = \{(e_1, 0), (e_2, 0.2), (e_3, 0)\} = P_{0.2}^{e_2}$, because
$f\big(F(e_1)\big) = f(\emptyset) = 0$, $f\big(F(e_2)\big) = f(\{x_2\}) = 0.2$ and $f\big(F(e_3)\big) = f(\emptyset) = 0$.

VII) The composite of fuzzy point P_α with soft point F_e^x, we have that: $P_\alpha \circ F_e^x = P_\alpha$

Example 2.2.9:

Let $X = \{x_1, x_2, x_3\}, E = \{e_1, e_2, e_3\}$
$P_{0.1} = \{(\{x_1\}, 0.1), (\{x_2\}, 0.1)(\{x_3\}, 0.1), (\{x_1, x_2\}, 0.1), (\{x_1, x_3\}, 0.1), (\{x_2, x_3\}, 0.1), (\emptyset, 0.1), (X, 0.1)\}$
$F_{e_2}^{x_2} = \{(e_1, \emptyset), (e_2, \{x_2\}), (e_3, \emptyset)\}$
Then $P_{0.1} \circ F_{e_2}^{x_2} = P_{0.1}$, because

$f(F(e_1)) = f(\emptyset) = 0.1$, $f(F(e_2)) = f(\{x_2\}) = 0.1$ and $f(F(e_3)) = f(\emptyset) = 0.1$
Thus $P_{0.1} \circ F_{e_2}^{x_2} = \{(e_1, 0.1), (e_2, 0.1), (e_3, 0.1)\} = P_{0.1}$

VIII) The composite of fuzzy point P_α with soft point F_e^x, we have that $P_\alpha \circ F^x = P_\alpha$

Example 2.2.10:

Let $F^{x_3} = [(e_1, \{x_3\}), (e_2, \{x_3\}), (e_3, \{x_3\})$ and $P_{0.1}$from example 2.2.9 we have : $P_{0.1} \circ F^{x_3} = P_{0.1}$ because : $f(F(e_1)) = f(F(e_2))= f(F(e_3)) = f(\{x_3\}) = 0.1$

IX) The composite of fuzzy point P_α with soft point F_e we have that $P_\alpha \circ F_e = P_\alpha$

Example 2.2.11:

Let F_{e_3} =$\{(e_1, \emptyset), (e_2, \emptyset), (e_3, \{x_1, x_3\})\}$ and $P_{0.1}$from example 2.2.9 then the composition : $P_{0.1} \circ F_{e_3} = P_{0.1}$, because
$f(F(e_1)) = f(F(e_2)) = f(\emptyset) = 0.1$ and $f(F(e_3)) = f(\{x_1, x_3\}) = 0.1$ thus :
$P_{0.1} \circ F_{e_3} = \{(e_1, 0.1), (e_2, 0.1), (e_3, 0.1)\} = P_{0.1}$.

Table (2-1) fuzzy soft points

Fuzzy points	Soft points	Fuzzy soft points
P_α^A	F^x with $A = \{x\}$	
P_α^C	F_e^x with $\emptyset, \{x\} \in C \subset P(X)$	
P_α^C	F^x with$\{x\} \in C$	
P_α^C	F_e with $\emptyset, F(e) \in C$	P_α
P_α	F_e^x	
P_α	F^x	
P_α	F_e	
P_α^A	F_e^x with $A = \{x\}$	
P_α^A	F_e with $A = F(e)$	
P_α^C	F_e^x with $\emptyset \notin C, \{x\} \in C$	P_α^e
P_α^C	F_e with $\emptyset \in C, F(e) \in C$	
P_α^C	F_e^x with $\emptyset \in C, \{x\} \notin C$	$P_\alpha^B, e \notin B \subset E$
P_α^C	F_e with $\emptyset \in C, F(e) \notin C$	

Conclusion 2.2.12 :-

1.Through the process of composite three fuzzy points with three kinds of soft points we get three different kinds of fuzzy soft points as showing in the above table 2-1.

2. Some new definitions and a new different types of fuzzy soft points via the composition between the three different fuzzy points presented in [26] with three different soft points mentioned in [36] with illustrative examples.

2.3:Composite Fuzzy Soft Topological Spaces:

In this section new definitions and theorems which concern with the composition between fuzzy topological spaces and soft topological spaces are presented to produce fuzzy soft topological spaces related with the fuzzy soft points mentioned in table (2.1) with defining base and sub base for spaces and different separation axioms with illustrative example are presented :

Definition 2.3.1:

Let X be universel set and E be the set of parameters of the elements of X , with $f: P(X) \to I$ and $F: E \to P(x)$,we call that the space $(\widetilde{1},\ {}_f\tau_s)$ be fuzzy soft topology with ${}_f\tau_s \subseteq I_{\widetilde{f}}^{E}$ if

i. $\widetilde{\emptyset},\ {}_fX_s \ \in\ {}_f\tau_s$

ii. If ${}_fA_s\,,\ {}_fB_s \in {}_f\tau_s \to {}_fA_s \wedge {}_fB_s \in {}_f\tau_s$

iii. For any index Γ and $\forall\lambda \in \Gamma$ let ${}_fA_{\lambda s} \in {}_f\tau_s \to \vee_{\lambda\in\Gamma}\ {}_fA_{\lambda s} \in {}_f\tau_s$

Definition 2.3.2:

The base of the fuzzy soft set ($\widetilde{1},\ {}_f\tau_s$) is the collection ${}_f\beta_s \subseteq {}_f\tau_s$ such that every members of ${}_f\tau_s$is the fuzzy union of some members of ${}_f\beta_s$ in others words ${}_f\beta_s$ is a base for ${}_f\tau_s$, that is

i. ${}_f\beta_s \subseteq {}_f\tau_s$

ii. For each $\widetilde{U} \in {}_f\tau_s$ and for each fuzzy point $\tilde{P}\ in\ \widetilde{U}$ there exists $\tilde{B}$ in ${}_f\beta_s$ such that $\tilde{P} \in \tilde{B} \leq \widetilde{U}$ where $\tilde{P}$ is the fuzzy point of type I or II or III

Definition 2.3.3 :

i. The collection of fuzzy soft sets ${}_f\beta_s^*$ is called sub base of the fuzzy soft topology ${}_f\tau_s$ if ${}_f\beta_s^* \subseteq {}_f\tau_s$

ii. the union of ${}_f\beta_s^*$ with all finite fuzzy intersection of members of ${}_f\beta_s^*$ is a base for ${}_f\tau_s$

Remarks 2.3.4:

Let X be universal set and E be the set of parameters of elements of X . The composite of fuzzy topology defined on $2^X \times I$ with soft topology defined on $E \times 2^X$, we get the base for fuzzy soft topology defined on $E \times I$

Example 2.3.5 :

Let X= $\{x_1, x_2, x_3\}$, E = $\{e_1, e_2, e_3\}$

$A_1 = \{e_1, e_2\}$ $B = \{e_2, e_3\}$, $C = \{e_2\}$

$F_{A_1} = \{ (e_1, \{x_1\}), (e_2, \{x_2, x_3\}), (e_3, \emptyset) \}$

$G_B = \{ (e_1, \emptyset), (e_2, \{x_1, x_2\}), (e_3, \{x_3\})\}$

$H_{1C} = \{ (e_1, \emptyset), (e_2, \{x_2\}), (e_3, \emptyset)\}$

$H_{2E} = \{ (e_1, \{x_1\}), (e_2, X), (e_3, \{x_3\})\}$

${}_s\tau = \{ \tilde{\tilde{\emptyset}}, \tilde{\tilde{X}}, F_A, G_B, H_{1C}, H_{2E}\}$

${}_f\tilde{A}_1 = \{ (\emptyset, 0), (\{x_1\}, 0.1), (\{x_2\}, 0), (\{x_3\}, 0), (\{x_1, x_2\}, 0.1),$ $(\{x_1, x_3\}, 0), (\{x_2, x_3\}, 0) (X, 0)\}$

And ${}_f\tau = \{ \tilde{0}, \tilde{1}, {}_f\tilde{A}_1 \}$

From the composite of ${}_s\tau$ *with* ${}_f\tau$ where :

$$\xrightarrow{\tilde{0}\text{ composite}} \begin{Bmatrix} \tilde{\emptyset} \\ \tilde{X} \\ F_{A_1} \\ G_B \\ H_{1c} \\ H_{2e} \end{Bmatrix} = f^{\tilde{0}}$$

$$\xrightarrow{\tilde{1}\text{ composite}} \begin{Bmatrix} \tilde{\emptyset} \\ \tilde{X} \\ F_{A_1} \\ G_B \\ H_{1_c} \\ H_{2_E} \end{Bmatrix} = f^{\tilde{1}}$$

$$\xrightarrow{\tilde{A}_1\text{composite}} \begin{cases} \tilde{\tilde{\emptyset}} = {}_f\tilde{0} \\ \tilde{\tilde{X}} = {}_f\tilde{0} \\ F_A = \{ (e_1, 0.1), (e_2, 0), (e_3, 0)\} = {}_fA_1A_S \\ \mathrm{G_B} = (e_1, 0), (e_2, 0.1), (e_3, 0)\} = {}_fA_1B_S \\ H_{1_c} = \{ (e_1, 0), (e_2, 0), (e_3, 0)\} = {}_1\tilde{0} \\ H_{2_E} = \{ (e_1, 0.1), (e_2, 0), (e_3, 0)\} = {}_fA_1E_S \end{cases}$$

Then the collection $\{ {}_f\tilde{0}, {}_f\tilde{1}, {}_fA_1A_S, {}_fA_1B_S\}$ is base

for ${}_f\tau_s = \{ {}_f\tilde{0}, {}_f\tilde{1}, {}_fA_1A_S, {}_fA_1B_S, {}_fA_1A_S \vee {}_fA_1B_S\}$

Note 2.3.6 : For any sets ${}_fA_S, {}_fB_S \in I_7^E$ is called fuzzy soft dis joint if ${}_fA_S \wedge {}_fB_S = 0$

- If ${}_fA_S$ is coming from composite of fuzzy set ${}_fA$ with soft set F_A and fuzzy soft ${}_fB_s$ is coming from composite of fuzzy set ${}_fB$ with soft set G_B, *so* ${}_fA_S \wedge {}_fB_S = {}_f\tilde{0}_S$ then it is not necessary that $F_A \cap \mathrm{G_B} = \emptyset$ *or* ${}_fA \wedge {}_fB = {}_f\tilde{0}$

$F_A \cap G_B = \emptyset$ *and* ${}_fA \wedge {}_fB = {}_f\tilde{0}$

Example 2.3.7 : Let $\mathbf{X} = \{x_1, x_2, x_3,\}$, $\mathbf{E} = \{e_1, e_2, e_3\}$

$F_A = \{ (e_1, \{x_1\}), (e_2, \emptyset), (e_3, \{x_2\}) \}$

G_B ={ $(e_1, \{x_2, x_3\})$, $(e_2, \{x_1, x_2\})$, $(e_3, \{x_1\})$}

${}_fC_1$={ $(\emptyset, 0.1)$, $(\{x_1\}, 0.1)$, $(\{x_2\}, 0.1)$,($\{x_3\}, 0)\}$,

($\{x_1, x_2\}, 0.2)$, $(\{x_2, x_3\}, 0.1)$, $\{x_1, x_3\}, 0.2)$, $(X, 0.2)$ }

${}_fC_2$={ $(\emptyset, 0)$, $(\{x_1\}, 0)$, $(\{x_2\}, 0.1)$,($\{x_3\}, 0)$,

($\{x_1, x_2\}, 0.1)$,($\{x_2, x_3\}, 0.1$), $\{x_1, x_3\}, 0.1)$, $(X, 0.2)$ }

${}_fC_3$={ $(\emptyset, 0.1)$, $(\{x_1\}, 0.1)$, $(\{x_2\}, 0)$,($\{x_1, x_2\}, 0)$,

($\{x_1, x_2\}, 0)$, $(\{x_2, x_3\}, 0)$, $(\{x_3\}, 0.1)$,(X,0)}

${}_fC_1 compostion \rightarrow F_A = \{ (e_1, 0.1), (e_2, 0.1), (e_3, 0.1)\}$

${}_fC_1 compostion \rightarrow \mathrm{G_B} = \{(e_1, 0.1), (e_2, 0.2), (e_3, 0.1)\}$

${}_fC_1 F_A \wedge {}_fC_1 G_{BS} \neq {}_f\tilde{0}$ and $F_A \cap G_B = \widetilde{\emptyset}$

${}_fC_2 compostion \rightarrow F_A = \{ (e_1, 0), (e_2, 0), (e_3, 0.1)\}$

${}_fC_3 compostion \rightarrow \mathrm{G_B} = \{ (e_1, 0.1), (e_2, 0.1), (e_3, 0)\}$

${}_fC_2 F_{A_S} \wedge {}_fC_3 F_{A_S} = {}_f\tilde{0}, but \ {}_fC_2 \wedge {}_fC_3 = {}_f\tilde{0}$ and

$F_A \cap F_A = F_A \neq \widetilde{\emptyset}$

${}_fC_3 compostion \rightarrow \mathrm{G_B} = \{(e_1, 0), (e_2, 0), (e_3, 0.1)\}$

${}_fC_2 F_{A_S} \wedge {}_fC_3 \mathrm{G}_{\mathrm{B}_S} \neq \tilde{0}, but \ {}_fC_2 \wedge {}_fC_3 = {}_f\tilde{0} \ and \ F_A \cap G_B = \widetilde{\emptyset}$.

Definition 2.3.8 :

A fuzzy soft set ${}_fA_S \in I_7^E$ is said to be fuzzy soft nhd of the composite fuzzy soft point $P(P\ is\ the\ tyPe\ I, II\ or\ III)$ iff there exists an fuzzy soft open set ${}_fU_S$ containing p and ${}_fU_S \leq {}_fA_S$. The collection of all fuzzy soft nhd of any fuzzy soft point P denoted by ${}_fN_s(P)$.

proposition 2.3.9 :

A fuzzy soft set ${}_fU_S$ is fuzzy soft open iff ${}_fU_S$ contains a fuzzy soft nhd of each its composite fuzzy soft points

noted that $i_\circ$ if ${}_fA_S$ fuzzy soft nhd of the composite fuzzy soft point $P_\propto^e$ .But $P_\propto^e$ coming from composite of :

1- A fuzzy point $P_\propto^B$ with soft point $F_e^x, A = \{x\}$,so ${}_fA_S$ is composite of fuzzy nhd of fuzzy points $P_\propto^A$ with soft nhd of soft point F_e^x

2- A fuzzy point $P_{\propto}^{A}$ with soft point F_e , $A = F_{(e)}$ so ${}_fA_S$ is composite of fuzzy nhd of fuzzy point $P_{\propto}^{A}$ with soft nhd of soft point F_e

3- A fuzzy point $P_{\propto}^{\mathcal{L}}$ with soft point F_e^x , $\emptyset \notin \mathcal{L}$ and $\{x\} \in \mathcal{L}$, so ${}_fA_S$ composite of fuzzy nhd of fuzzy point $P_{\propto}^{\mathcal{L}}$ with soft nhd of soft point F_e^x

4- i- A fuzzy point $P_{\propto}^{\mathcal{L}}$ with soft point F_e , $\emptyset \notin \mathcal{L}$, $F(e) \tilde{\in} \mathcal{L}$. So ${}_fA_S$ is composite of fuzzy nhd of fuzzy point $P_{\propto}^{\mathcal{L}}$ with soft nhd of soft point F_e.

ii- Also if ${}_fA_S$ is fuzzy soft nhd of the composite fuzzy soft point of P_α , $P_{\propto}^{B}$ *and* $e \notin B \subset E$ we can use the table 2-1 of composite fuzzy points with soft points in order to analyses the fuzzy soft ${}_fA_S$ as was released in the first part

iii- if ${}_fA$ is fuzzy nhd of fuzzy point $P_{\propto}^{\{x\}}$ $(P_\alpha, P_{\propto}^{\mathcal{L}})$ and F_B ,is soft nhd of soft point F_e^x (F^x, F_e) then fuzzy soft set ${}_fA_S$(which is coming from the composite of fuzzy set ${}_f\tilde{A}$ with soft set F_B is fuzzy soft of composite fuzzy soft point $P_{\propto}^{e}$.

Proof : since ${}_fA$ is fuzzy nhd of fuzzy point $P_x^{\propto}$,then there exists fuzzy open set ${}_fU$ such that $P_x^{\alpha} \tilde{\in} {}_fU \leq {}_fA$ .Also F_B is soft nhd of soft point F_e^x ,there exist soft open set that $G_{B_1} s.t.\ F_e^x \in G_{B_1} \subseteq F_B$

But the composite of fuzzy open ${}_fU$ with soft open G_{B_1} is fuzzy soft basic open set ${}_fU_{B_{1s}}$ and the composite of fuzzy point $F_{\propto}^{\{x\}}$ with soft point P_e^x is fuzzy soft point $P_{\propto}^{e}$ .But the composite of fuzzy set ${}_fA$ with soft set F_A is fuzzy soft set ${}_fA_S$ there for $P_{\propto}^{e} \in {}_fU_{B1S} \leq {}_fA_S$ Hence ${}_fA_S$ is fuzzy soft nhd of the composite fuzzy soft point $P_{\propto}^{e}$

Note that by similar way we take the fuzzy points $P_{\propto}$ *or* $P_{\propto}^{\mathcal{L}}$ with soft points F^x, F_e and we use that table of composite fuzzy points with soft points to will it .

Defintion.2.3.10.

The fuzzy soft topology ($\tilde{1}$, ${}_f\tau_S$) is called

i. Fuzzy soft $T_\circ$- space iff for any two different composite fuzzy soft point of the same kinds or different kinds there exist fuzzy soft open set contains one of them not the other .

ii. Fuzzy soft T_1- space iff for any two different composite fuzzy soft point of the same kinds or different kinds there exists two fuzzy soft open sets each one of them contains one of these points and not the anther.

iii. Fuzzy soft T_2- space iff for any two different composite fuzzy soft point of the same kinds or different kinds there exists two disjoint fuzzy soft open sets each of them contains one of them not the other .

iv. Fuzzy soft regular space iff for any composite fuzzy point and fuzzy soft closed set not containing it, there exists two disjoint fuzzy soft open sets one of them contains the point and the other contains the fuzzy soft closed set.

Now let that for any composite fuzzy soft point it is the result of the composite of fuzzy point with soft point as shown in the table(2-1). So for any two different composite fuzzy soft point of the kind P_α depends on the value of the membership function of the fuzzy soft points this is turn comes from the difference values of the membership functions of the fuzzy point of any kinds in processing of the composition.

But for any two different composite fuzzy points of the kind P_α^e depends on the value of the membership of functions of fuzzy soft point and the value of e of parameters of the numbers of the universe set X ,this in turn comes from the values of membership functions of the fuzzy points of the kind P_α^A , $P_\alpha^\mathcal{L}$ with soft fuzzy points of the kinds F_e on F_e^x as shown in the table(2-1)

Finally for any two different composite fuzzy soft points of the type P_α^B , B$\subset E$ depends on the value α and B this in turn comes from the values of membership function of the fuzzy point of type $P_\alpha^\mathcal{L}$ with the soft points of types F_e^x , F_e and conditions that e$\notin$B , $\emptyset\epsilon\mathcal{L}$, $\{x\} \notin \mathcal{L}$ and $F(e) \notin \mathcal{L}$ see table 2-1 through which we will get the following results.

Theorem 2.3.11 :

If a fuzzy soft topology (${}_f\tilde{1}$, ${}_f\tau_s$) is a fuzzy soft $T_\circ$- space with respect to the composite point of kinds $P_\propto^e$ and $P_\propto^B$, then fuzzy topology($\tilde{1}$, ${}_f\tau$).is fuzzy $T_\circ$- space with aspect to fuzzy point $P_\propto^A$ and $P_\propto^\mathcal{L}$

Proof. Suppose that the $fSt\ ({}_f\tilde{1},\ {}_f\tau_s)$ is $fsT_\circ$- space and let $P^A_{\propto 1} \neq P^A_{\propto 2}$, so the composite of $P^A_{\propto 1} and P^A_{\propto 2}$ with any soft point of kinds $F^x_e or\ F_e$ with $A = \{x\}$ or $A = F_{(e)}$, we get two different composite fuzzy soft point $P^e_{\alpha 1} \neq P^e_{\alpha 2}$, so there exist fuzzy soft open set ${}_fU_s$ containing $P^e_{\alpha 1}$ but not $P^e_{\alpha 2}$or containing $P^e_{\alpha 2}$, but not $P^e_{\alpha 1}$ .However ${}_fU_s$ is composite of fuzzy open with soft open thus there exists fuzzy open set $\tilde{A}$ and soft open set F_A containing $F^\alpha_e (F_e) s.t. \overrightarrow{\tilde{A}\ compostion}\ F_A = {}_fU_s$ and $\tilde{A}$ containing $P^A_{\propto 1} not P^A_{\propto 2}$ or containing $P^A_{\propto 2}$ but not $P^A_{\propto 1}$ .Similarly the other case $P^{A_1}_{\propto 1} \neq P^{A_2}_{\propto 2}$ $(A_1 = \{x_1\} and\ A_2 = \{x_2\}$ or $A_1 = F_{(e1)}\ and\ (A_2 = \mathrm{F}(\mathrm{e}_2))$

Theorem 2.3.12:

If the fuzzy topology $(\tilde{1},\ {}_f\tau\ )$ is $fT_\circ - space$ with respect to $P^A_\propto$, $P^{\mathcal{L}}_\propto$and the soft topology$(\tilde{X},\ {}_s\tau)$ is $ST_\circ - space$ with respect to F^x_e , F_e , then the composite fuzzy soft topology $({}_f\tilde{1},\ {}_f\tau_s)$ is $f\ {}_sT_\circ - space$ with respect to composite fuzzy soft point $P^e_\propto$

proof :

let $P^e_{\propto 1} \neq P^e_{\propto 2}$or $P^{e_1}_{\propto_1} \neq P^{e_2}_{\propto_2}$ so we get The following cases of the compositions. Case I

1- $P^{e_1}_{\propto_1} \neq P^{e_2}_{\propto_2} compostion\ \rightarrow P^A_{\propto_1} \neq P^A_{\propto_2}\ with\ F_e$, A= F(e)

composite $\rightarrow$ $.P^{\mathcal{L}}_{\propto_1} \neq P^{\mathcal{L}}_{\propto_2}\ with\ F_e\ , \emptyset \notin \mathcal{L}$

And $\{x_1\} \in \mathcal{L}$

composite $\rightarrow$ $P^{\mathcal{L}}_{\propto_1} \neq P^{\mathcal{L}}_{\propto_2}\ with\ F_e\ , \emptyset \in \mathcal{L}$

And

$F(e)\ \notin \mathcal{L}$

case II:

$P_{\propto_1}^{e_1} \neq P_{\propto_2}^{e_2}$ $\xrightarrow{\text{composite}}$ $P_{\propto_1}^{A} \neq P_{\propto_2}^{A}$ with $F_{e_1}^{x} \neq F_{e_2}^{x}, \{x\} = A$

$\xrightarrow{\text{composite}}$ $P_{\propto_1}^{A_1} \neq P_{\propto_2}^{A_2}$ with $F_{e_1} \neq F_{e_2}$, with, $A_1 = F(e_1), A_2 = F(e_2)$

$\xrightarrow{\text{composite}}$ $P_{\propto_1}^{A_1} \neq P_{\propto_2}^{A_2}$ with $F_{e_1}^{x_1} \neq F_{e_2}^{x_2}, A_1 = \{x_1\}$ and $A_2 = \{x_2\}$

case III :

$\neq p_{\propto_2}^{e_2}$

$\xrightarrow{\text{composite}}$ $P_{\propto_1}^{\mathcal{L}} \neq P_{\propto_2}^{\mathcal{L}}$ with $F_{e_1}^{x} \neq F_{e_2}^{x}$, and $\emptyset \notin \mathcal{L}$; $\{x\} \in \mathcal{L}$

$\xrightarrow{\text{composite}}$ $P_{\propto_1}^{\mathcal{L}} \neq P_{\propto_2}^{\mathcal{L}}$ with $F_{e_1}^{x_1} \neq F_{e_2}^{x_2}$, and $\emptyset \notin \mathcal{L}$; $\{x_1\}, \{x_2\} \in \mathcal{L}$

$\xrightarrow{\text{composite}}$ $P_{\propto_1}^{\mathcal{L}_1} \neq P_{\propto_2}^{\mathcal{L}_2}$ with $F_{e1}^{x1} \neq F_{e2}^{x2}$ and $\emptyset \notin \mathcal{L}_1, \mathcal{L}_2$ and $\mathcal{L}_1, \mathcal{L}_2$ and $\mathcal{L}_1, \mathcal{L}_2$ are disjoint also $\{x_1\} \in \mathcal{L}_1, \{x_2\} \in \mathcal{L}_2$

$\xrightarrow{\text{composite}}$ $P_{\propto_1}^{\mathcal{L}} \neq P_{\propto_2}^{\mathcal{L}}$ with $F_{e1} \neq F_{e2}$ and $\emptyset, F(e_1), F(e_2) \in \mathcal{L}$

$\xrightarrow{\text{composite}}$ $P_{\propto_1}^{\mathcal{L}_1} \neq P_{\propto_2}^{\mathcal{L}_2}$ with $F_{e1} \neq F_{e2}$ and $\emptyset$, $F(e_1) \in \mathcal{L}_1, \emptyset, F(e_2) \in \mathcal{L}_2$ and $\mathcal{L}_1 \cap \mathcal{L}_2 = \emptyset$

$\xrightarrow{\text{composite}}$ $P_{\propto_1}^{\mathcal{L}_1} \neq P_{\propto_2}^{\mathcal{L}_2}$ with $F_{e_1}^{x} \neq F_{e2}$ and $\emptyset \notin \mathcal{L}_1$, $\{x\} \in \mathcal{L}_1$, and $\emptyset \in \mathcal{L}_2$ $F(e_2) \in \mathcal{L}_2$

$\xrightarrow{\text{composite}}$ $P_{\propto_1}^{A} \neq P_{\propto_2}^{\mathcal{L}}$ with $F_{e1}^{x} \neq F_{e2}$ and $A = \{x\}$, $\emptyset \in \mathcal{L}$ and $F(e_2) \in \mathcal{L}$

From these cases we get different fuzzy point and soft point ,but *by* assumption that fuzzy space($_f\tilde{1}$, $_f\tau$) is $fT_\circ$- space and soft space ($\tilde{x}$, $_S\tau$) is $sT_\circ$- space.

There exist fuzzy open set $\tilde{A}$ contains one of the different two fuzzy point (as shown in the cases) not the other and soft open set F_A contains of the other different two soft point (as shown in the cases) not others . Hence the composite of $\tilde{A}$ with F_{A1} we get the composite open basic

$AA\ \overrightarrow{composite}$ $\tilde{A}$ with F_{A1} containing one of the two different composite fuzzy soft point not the others . therefore the $({}_f\tilde{1},\ {}_f\tau_s\ )\ is\ fsT_{\circ}$- space.

Theorem 2.3.13:

The fuzzy soft topology $({}_f\tilde{1}$ is $fsT_{\circ}$- space with respect to composite fuzzy soft point of kind if the fuzzy topology ($\tilde{1}$, is - space with respect to fuzzy point and the soft topology ($\tilde{X}$, is - space w.r.t. soft point

Proof: Since for any two different composite fuzzy soft point dependence on the values of so we have the following different cases by using the table(2-1)

case I

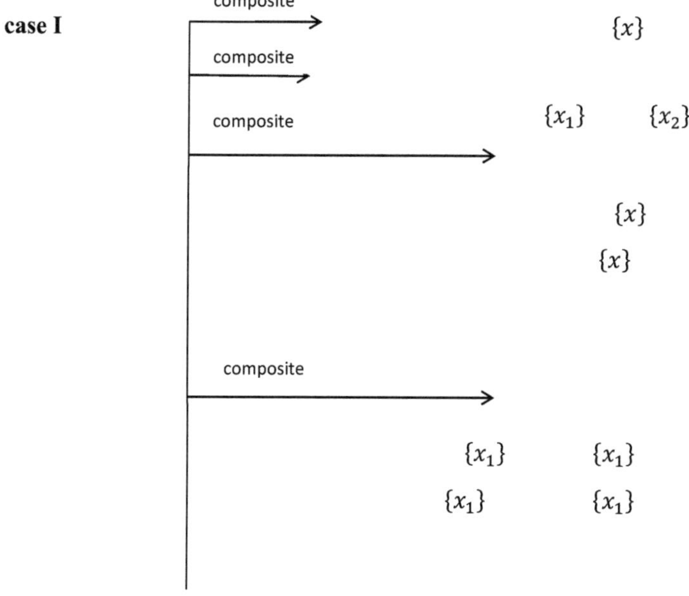

case II

Composite

$\{x\}$

$\{x_1\}$, $\{x_2\}$

$\{x_1\}$, $\{x_1\}$

$\{x\}$ $\mathcal{L}$.

Case III

composite

1) F $\{x_1\}$

2) F $\{x_1\}$

3)

$\{x_1\}$

composite

$\{x_1\}$

$\{x_1\}$

$\{x_1\}$

So for any two different composite fuzzy soft points there exists a fuzzy open set $\tilde{A}$ containing one of the other fuzzy points is included in the installation composite not the other also there exist soft open set contains the soft points $(F^x, F_e\, or F_e^x)$ or contains one of the other soft points in the case that exist two different soft points that included in the installation of composite not the other . Therefore the composite of the fuzzy open with soft open we get the fuzzy soft basic one containing P_{α_1}not P_{α_2} or contains P_{α_2}not P_{α_1}.

Theorem 2.3.14: If the fuzzy soft topology $({}_f\tilde{1},{}_f\tau_s)$is f S $T_\circ$-space with respect to composite fuzzy soft point P_α, then the fuzzy topology $(\tilde{1}.{}_f\tau)$ is f$T_\circ - sPace$ with respect to fuzzy points $P_\alpha^A, P_\alpha^{\mathcal{L}}\ and\ P_\alpha$.

Proof: If we take two different fuzzy points (on the same kinds or different), and we composite with any soft point (of any kind) we get the two different composite fuzzy point of kind P_α , so there exists composite fuzzy soft open set containing one of them not the other ,then the fuzzy open set included in the installation composite contains one of the different fuzzy point not the other. The same technique used in the proofs of theorems 2.3.11 , 2.3.12 , 2.3.13 and 2.3.14 we can the following theorems.

Theorem 2.3.15:

If a fuzzy soft topology $({}_f\tilde{1}, {}_f\tau_s)$ is a fuzzy soft T_1 −space with respect to composite fuzzy soft point of kinds $P_\alpha^{\mathcal{L}}, P_\alpha^B$ then the fuzzy topology $(\tilde{1},{}_f\tau)$ is a fuzzy T_1 −space with respect to fuzzy point of kind $P_\alpha^A\ and\ P_\alpha^{\mathcal{L}}$.

Theorem 2.3.16:

If a fuzzy topology $(\tilde{1}, {}_f\tau)$ is a fuzzy soft f T_1 −space with respect to fuzzy points $P_\alpha^A, P_\alpha^{\mathcal{L}}$ and soft topology $(\tilde{X}, {}_S\tau)$ is S T_1 −space with respect to the soft points $F_e and\ F_e^x$, then the composite fuzzy soft topology $({}_f\tilde{1}, {}_f\tau_s)$ is f S T_1 −space with respect to fuzzy soft point P_α

Theorem 2.3.17:

If the fuzzy topology $(\tilde{1}, {}_f\tau)$ is f T_1 −space with respect to fuzzy points P_α $P_\alpha^A and P_\alpha^\ell$ and soft topology $(\tilde{X}, {}_S\tau)$ is S T_1 −space with respect to the soft points F_e , $F^x and\ F_e^x$, then the composite fuzzy soft topology $({}_f\tilde{1}, {}_f\tau_s)$ is f S T_1 −space with respect to composite fuzzy soft pointP_α^e.

Theorem 2.3.18: If the fuzzy soft topology $({}_f\tilde{1}, {}_f\tau_s)$ is f S T_1 −space with respect to composite fuzzy soft point P_α, then the fuzzy topology $(\tilde{1}, {}_f\tau)$ is f T_1space with respect to fuzzy points $P_\alpha, P_\alpha^A and\ P_\alpha^{\mathcal{L}}$.

Definition2.3.19 : A fuzzy topology $(1, {}_f\tau)$ is called fuzzy compact iff every fuzzy open cover has finite fuzzy open cover.

Theorem 2.3.20: if the fuzzy soft topology $({}_f\tilde{1}, {}_f\tau_s)$ is fuzzy soft compact , then the fuzzy space $(\tilde{1}, {}_f\tau)$ and soft topology $(\tilde{X}, {}_S\tau)$ is fuzzy compact and soft compact respectively.

Proof : Let G_f be a fuzzy open cover for fuzzy space $(\tilde{1}, {}_f\tau)$ and Gs be soft open cover for the soft space $(\tilde{X}, {}_S\tau)$, Then the possibilities of composite fuzzy open sets in the cover G_f with soft open sets in the cover Gs, we get fuzzy soft open cover $f^G s$ the fuzzy soft space $({}_f\tilde{1}, {}_f\tau_s)$ but $({}_f\tilde{1}, {}_f\tau_s)$is fuzzy soft compact, then there exists finite fuzzy soft open cover to the space $({}_f\tilde{1}, {}_f\tau_s)$.Therefore the fuzzy open set and the soft set that in the installation composite of finite fuzzy soft open cover to the space $(\tilde{1}, {}_f\tau)$ as finite soft open cover to the soft space $(\tilde{X}, {}_S\tau)$.

Hence the space$(\tilde{1}, {}_f\tau)$ is fuzzy compact and the space $(\tilde{X}, {}_S\tau)$.is soft compact.

Exercises

1- Let X = { x , y , z } , E = { e , a } , F(e) = {x,y } , F(a)= { x }

F_E = { (e , F (e)) , (a , F (a)) } and $_fA$ = { ({x} , 0), ({y}, 0.1) , ({z} , 0) , ({x,y }, 0.01) , ({x,z}, 0.2) , ({y,z}, 0.4), (∅ ,0), (X, 0.7) } . Find the following

a- The composite of $_fA$ with F_E

b- The composite of $P^{\{y,z\}}{}_{0.3}$ with soft point F_a^z

c- The composite of $P_{0.2}{}^{\{x,y\}}$ with soft point F_e where F(e)= {z}

d- The composite of fuzzy point $P_{0.1}{}^X$ with soft point F_y .

2- Given an example explain that the composite of fuzzy point $P_\beta{}^C$ with soft point F_e with $\emptyset \in C , F(e) \notin C \subset P(X)$.

3- Given an example explain that the composite of fuzzy point $P^C{}_\beta$ with soft point F_e^x is fuzzy point $P_\beta{}^e$.

4- Let X = { x, y } , E = { 1 , 2 , 3 } , A ={ 1 , 2 },B = {2}

F_A = { (1 , X) , (2 , { x}) , (3 , ∅) }

G_B = { (1 , ∅), (2 ,{ y }) , (3 , ∅)}

H_A = { (1, X) , (2 , X) , (3 , ∅) }

$\tau_S = \{ \emptyset_S , X_S , F_A , G_B , H_A \}$

$_f\tau = \{ {}_f0, {}_f1 , {}_fC\}$, $_fC$ = { (∅, 0.1) ,({x} , 0.7),({y}, 0.8), (X, 0.1)}

Find fuzzy soft topology $_f\tau_S$ and given two different basis of this topology .

5- Show that the collection of fuzzy soft neighborhoods to fuzzy soft point $P_\propto$ satisfy the following

a- If $_fA_S \in {}_fN_S(P_\propto)$ *and* $_fA_S \leq {}_fB_S$, then $_fB_S \in {}_fN_S(P_\propto)$

b- If ${}_fA_s \in {}_fN_s(P_\propto)$, $then\ P_\propto \leq {}_fA_s$

c- If ${}_fA_s$, ${}_fB_s \in {}_fN_s(P_\alpha)$, $then\ ({}_fA_s \wedge {}_fB_s) \in {}_fN_s(P_\alpha)$

6- Let ${}_f\beta_s$ be a base of fuzzy soft topology ${}_f\tau_s$ .Show that for each

$$ {}_fU_s \in {}_f\tau_s , \exists\, \Gamma \subseteq {}_f\beta_s\ s.t.\ {}_fU_s = \bigcup \Gamma $$

7- Define the interior , exterior , boundary , accumulation and adherent fuzzy soft points similarly as the in fuzzy topological spaces and state the properties of these points with proves and examples explain these definitions .

8- All the concepts in fuzzy topological spaces can states in fuzzy soft topological space , so explain the following

a- If The fuzzy soft set ${}_fA_s$ fuzzy soft dense , then fuzzy set and the soft set that interior in composite are fuzzy dense and soft dense respectively .Is the opposite true prove that , if not given an example explain that .

b- If the fuzzy set ${}_fA$ is fuzzy some -where dense and soft set G_B is soft some -where dense , then the composite fuzzy soft set ${}_fA_s$ is fuzzy soft some -where dense .

9- The fuzzy soft set is fuzzy soft connected if the fuzzy set and soft set that enter in the composite are fuzzy connected and soft connected respectively .

Chapter Three

Soft Fuzzy Sets and Soft Fuzzy Topological Spaces

3. Preliminaries

In this chapter the definition of soft fuzzy set with respect to the seventh family of fuzzy sets as a more general one are reformulated . Via the combination between the fuzzy points introduced by [26] with the soft points which is related to the first soft family $SS_1(x)$ presented in [34], we obtain a new five different kinds of soft fuzzy points and via the combination between the soft topology and fuzzy topology have obtained. A new soft fuzzy topological spaces and define base and subbase for a soft fuzzy spaces and a new soft fuzzy separation axioms related with new soft fuzzy points with some theorems and appended with illustrative examples are produced.

3.1. The Combination Between the Soft Sets and Fuzzy Sets

In this section some definitions are introduced and a new five soft fuzzy points are produced via the combination between three types of fuzzy points and three types of soft points .

The definition of Yao et al [67]is reformulated for a soft fuzzy set as follows:

Definition 3.1.1:

Let X be an initial universe set and E be a set of parameters. Let P(X) denotes the power set of X, the soft fuzzy set denoted by ${}_S F_f$ *is*

${}_S F_f$= {(e , ${}_f$F(e)) , $\forall e \in E$, ${}_f$F(e) $\in I_7^x$ } where ${}_f$F: $A \rightarrow I_7^x$

$$F: A \rightarrow P(X), \qquad \forall A \subseteq E$$

${}_f$F(e) = $\{\left(x, f_{F(e)}(x)\right), x \in X, f: X \rightarrow I\}$ also

$f_{F(e)}(x) = f(x), \forall x \in F(e)$ and $f_{F(e)}(x) = 0 , \forall x \notin F(e)$

We denote that the null soft fuzzy set and absolute soft fuzzy set are respectively as follows:

$\tilde{\tilde{\emptyset}} = \{(e, \tilde{0}), \forall e \in E\}$ is the null sot fuzzy set.

$\tilde{\tilde{X}} = \{(e, \tilde{1}), \forall e \in E\}$ is the absolute soft fuzzy set.

Note that if $F(a) = \emptyset$ then $f_{\emptyset}(x) = 0, \forall x \in X$, because $x \notin \emptyset , \forall x \in X$ and the family of all soft fuzzy sets denoted by $S(I_7^U)$

Example 3.1.2:

Let $f : X \rightarrow I, X = \{x_1, x_2, x_3, x_4, x_5\}, E = \{e_1, e_2, e_3, e_4\}$

$$f(x_1)=0.2 , f(x_2)=0.9, f(x_3)=0.4 ,f(x_4)=0.5 , f(x_5)=0.8$$

$$\tilde{A} = \{(x_1, 0.2), (x_2, 0.9), (x_3, 0.4), (x_4, 0.5), (x_5, 0.8)\}$$

is a fuzzy set and F: E→$P(X)$

$$\left.\begin{array}{l} F(e_1) = \{x_2, x_4\} \\ F(e_2) = X \\ F(e_3) = \phi \\ F(e_4) = \{x_1, x_3, x_5\} \end{array}\right\}$$

$F_A = \{(e_1, F(e_1)), (e_2, F(e_2)), (e_3, F(e_3)), (e_4, F(e_4))\}$ is a soft set

$f_{\{x_2,x_4\}} = f_{F(e_1)} = \{(x_1, 0), (x_2, 0.9), (x_3, 0), (x_4, 0.5), (x_5, 0)\}$

$f_X = f_{F(e_2)} = \{(x_1, 0.2), (x_2, 0.9), (x_3, 0.4), (x_4, 0.5)(x_5, 0.8)\}$

$f_{F(e_3)} = \{(x, 0), \forall x \in X\} = \tilde{0}$

$f_{F(e_4)} = \{(x_1, 0.2), (x_2, 0), (x_3, 0.4), (x_4, 0)(x_5, 0.8)\}$

Thus ${}_S F_f$=$\{(e_1, f_{F(e_1)}), (e_2, f_{F(e_2)}), (e_3, f_{F(e_3)}), (e_4, f_{F(e_4)})\}$ soft fuzzy set.

Definition 3.1.3:

A soft fuzzy set${}_S A_f$ is said to be soft fuzzy subsets of a soft fuzzy set ${}_S B_f$ over common universe set X iff

i. $A \subseteq B$ ii. $F(a) \subseteq G(a), \forall a \in A$ iii. $f_{F(a)}(x) \leq g_{G(a)}(x)$ $, \forall x \in X.$

We noted that from the combination of soft set F_A with fuzzy set $\tilde{A}$ we get the soft fuzzy set ${}_S A_f$where f is the membership of the fuzzy set $\tilde{A}$. Also the combination of soft set G_B with fuzzy set $\tilde{B}$ we get the soft fuzzy set ${}_S B_f$ where g is the membership of the fuzzy set $\tilde{B}$.

Definition 3.1.4:

Let ${}_S A_f$ and ${}_S B_f$ are two soft fuzzy sets then

i. The soft fuzzy intersection of ${}_S A_f$and ${}_S B_f$ denoted by ${}_S A_f \cap {}_S B_f = {}_S C_f$, where $H_C = F_A \cap G_B$ and $h(x) = \min_{x \in X}(f(x), g(x))$ with $H(a) = F(a) \cap G(a), \forall a \in A \cup B \subseteq E.$

Through this definition we see that the combination of the soft intersection for F_A and G_B with the fuzzy intersection for $\tilde{A}$ and $\tilde{B}$ respectively we have the soft fuzzy set ${}_SC_f$.

ii. The soft fuzzy union of ${}_SA_f$ and ${}_SB_f$ denoted by ${}_SA_f \cup {}_SB_f = {}_SC_f$ where $H_C = F_A \cup G_B$ and $h(x) = \max_{x \in X}(f(x), g(x))$ with $H(a) = F(a) \cup G(a)$,
$\forall a \in A \cup B \subseteq E$.
Also through this definition we see that the combination of the soft union for F_A and G_B with the fuzzy union for $\tilde{A}$ and $\tilde{B}$ respectively we have the soft fuzzy set ${}_SC_f$.

Example 3.1.5:
Let $X = \{x_1, x_2, x_3\}, E = \{e_1, e_2, e_3, e_4\}, A = \{e_1, e_2\}, B = \{e_1, e_2, e_3\}, C = \{e_2, e_3\}, F_A = \{(e_1, \{x_1\}), (e_2, \{x_1, x_2\}), (e_3, \emptyset), (e_4, \emptyset)\}, G_B = \{(e_1, \{x_1\}), (e_2, \{x_1, x_2\}), (e_3, \{x_3\}), (e_4, \emptyset)\}, U_C = \{(e_1, \emptyset), (e_2, \{x_1\}), (e_3, \{x_3\}), (e_4, \emptyset)\}$.
The fuzzy set is

$$\tilde{A} = \{(x_1, 0.1), (x_2, 0), (x_3, 0.3)\},$$
$$\tilde{B} = \{(x_1, 0.3), (x_2, 0.1), (x_3, 0.4)\},$$
$$\tilde{C} = \{(x_1, 0.2), (x_2, 0), (x_3, 0.33)\}.$$

Then

$${}_SA_f = \{(e_1, p_{0.1}^{x_1}), (e_2, p_{0.1}^{x_1}), (e_3, \tilde{0}), (e_4, \tilde{0})\},$$
$${}_SB_f = \{(e_1, p_{0.3}^{x_1}), (e_2, \{(x_1, 0.3), (x_2, 0.1), (x_3, 0)\}), (e_3, p_{0.1}^{x_2}), (e_4, \tilde{0})\}$$
$${}_SC_f = \{(e_1, \tilde{0}), (e_2, p_{0.2}^{x_1}), (e_3, p_{0.33}^{x_3}), (e_4, \tilde{0})\}.$$

We see that ${}_SA_f \subseteq {}_SB_f$.

$${}_SB_f \cup {}_SC_f = \left\{ \begin{matrix} (e_1, p_{0.3}^{x_1}), (e_2, \{(x_1, 0.3), (x_2, 0.1), (x_3, 0)\}), \\ (e_3, \{(x_1, 0), (x_2, 0.1), (x_3, 0.33)\}), (e_4, \tilde{0}) \end{matrix} \right\},$$
$${}_SA_f \cap {}_SC_f = \{(e_1, \tilde{0}), (e_2, p_{0.2}^{x_1}), (e_3, 0), (e_4, \tilde{0})\}.$$

Proposition 3.1.6:

Let X be initial universe set, E be set of a parameters of elements of the set X. For any soft fuzzy sets ${}_{S}A_f$ and ${}_{S}B_f$, if ${}_{S}A_f \cap {}_{S}B_f = \tilde{\tilde{0}}$ iff one of the following statement is hold:

1) $F_A \cap G_B = \tilde{\varphi}$
2) $\tilde{A} \cap \tilde{B} = \tilde{0}$
3) $F_A \cap G_B = \tilde{\tilde{0}}$ and $\tilde{A} \cap \tilde{B} = \tilde{0}$

Proof. Necessity, assume that ${}_{S}A_f \cap {}_{S}B_f = \tilde{\tilde{0}}$, then we get one of the statements 1,2 or 3 is hold from definition 3.1.4 i .

Sufficiency:

1. if $F_A \cap G_B = \tilde{\varphi}$, we get that $H(e) = F(e) \cap G(e) = \emptyset, \forall e \in E$ implies that ${}_{S}A_f \cap {}_{S}B_f = \tilde{\tilde{0}}$
2. if $\tilde{A} \cap \tilde{B} = \tilde{0}$, we get that $h(x) = min(f(x), g(x)) = 0, \forall x \in X$ implies that ${}_{S}A_f \cap {}_{S}B_f = \tilde{\tilde{0}}$
3. Also, if $F_A \cap G_B = \tilde{\varphi}$ and $\tilde{A} \cap \tilde{B} = \tilde{0}$, implies that

${}_{S}A_f \cap {}_{S}B_f = \tilde{\tilde{0}}$.

Note 3.1.7 : We belief from the definition 3.1.1 that soft fuzzy set is the final form and not fuzzy soft set since in the final form the set appears soft set and not fuzzy set so our belief is(soft fuzzy) hence the points are said to be soft fuzzy points and then via the combination of three different types of fuzzy points and three different types of soft points we get five different types soft fuzzy points as in section below .

3.2 Extraction the Soft Fuzzy Points

From the combination between three different types of fuzzy points with three different types of soft points we get the following types of soft fuzzy points .

Definition 3.2.1:

I. Let F_e^x be a soft point and p_α^x be a fuzzy point ,then the soft fuzzy point which is the combination of the soft point of type F_e^x with fuzzy point of type p_α^x we get soft fuzzy point of type

$$F_{P_\alpha^x}^e \ni F_{P_\alpha^x}^e(\varepsilon)=\begin{cases} P_\alpha^x & if\ \varepsilon = e \\ \tilde{O} & if\ \varepsilon \neq e \end{cases}$$

Example 3.2.2:

Let $\mathbf{X}=\{x_1, x_2, x_3, x_4, x_5\}$,E $=\{e_1, e_2, e_3, e_4\}$ and $F_{e_1}{}^{x_1}=\{(e_1, \{x_1\}), (e_2, \emptyset), (e_3, \emptyset), (e_4, \emptyset)\}$

$P_{0.2}^{x_1}=\{(x_1, 0.2), (x_2, 0), (x_3, 0), (x_4, 0), (x_5, 0)\}$

The membership function of soft fuzzy point $F_{P_\alpha^{x_1}}^{e_1}$

$f_{F(e_1)} = f_{\{x_1\}} = \{(x_1, 0.2), (x_2, 0), (x_3, 0), (x_4, 0), (x_5, 0)\} = P_{0.2}^{x_1}$

$f_{F(e_2)} = f_\emptyset = \tilde{o}$

$f_{F(e_3)} = f_{F(e_4)} = f_\emptyset = \tilde{o}$

$F_{P_\alpha^{x_1}}^{e_1}=\{(e_1, P_{0.2}^{x_1}, (e_2, \tilde{o}), (e_3, \tilde{o}), (e_4, \tilde{o})\}$.

NOTE 3.2.3:

Let $F_e{}^x$ be the soft point and P_α^y be a fuzzy point if $x \neq y$, then the soft fuzzy point $F_{P_\alpha^y}^x = \tilde{\tilde{o}}$, because $f_{F(e)} = \tilde{o}, \forall\ e \in E$ as shown in example below:

Example 3.2.4 :

Let X $=\{x_1, x_2, x_3, x_4, x_5\}$, E $=\{e_1, e_2, e_3, e_4\}$and

$F_{e_1}{}^{x_1}=\{(e_1, \{x_1\}), (e_2, \emptyset), (e_3, \emptyset), (e_4, \emptyset)\}$,

$P_{0.2}^{x_2}=\{(x_1, 0), (x_2, 0.2), (x_3, 0), (x_4, 0), (x_5, 0)\}$,

$f_{F(e_1)} = f_{F(e_2)} = f_{F(e_3)} = f_{F(e_4)} = \tilde{o}$.

Thus $F_{P_\alpha^{x_2}}^{e_1} = \tilde{\tilde{o}}$

II. The combination of the soft point of type F^x with fuzzy point of type P_α^x for some $x \in X$ and $0 < \alpha < 1$, we get the soft fuzzy point of type $F_{P_\alpha^x}$, that is $F_{P_\alpha^x}=\{(\mathrm{e}, P_\alpha^x) , \mathrm{e} \in E\}$.

Example 3.3.5:

Let X = $\{x_1, x_2, x_3, x_4, x_5\}$, F_{x_1}=$\{(e, \{x_1\}), e \in E\}$

$P_{0.2}^{x_1}$ =$\{(x_1, 0.2), (x_2, 0), (x_3, 0), (x_4, 0), (x_5, 0)\}$

$f_{F(e_1)}$ =$f_{\{x_1\}}$ =$\{(x_1, 0.2), (x_2, 0), (x_3, 0), (x_4, 0), (x_5, 0)\}$= $P_{0.2}^{x_1}$

$f_{F(e_2)}$ =$f_{\{x_1\}}$ =$\{(x_1, 0.2), (x_2, 0), (x_3, 0), (x_4, 0), (x_5, 0)\}$= $P_{0.2}^{x_1}$

$f_{F(e_3)}$=$f_{F(e_4)}$ = $f_{\{x_1\}}$ = $P_{0.2}^{x_1}$ =$\{(e, P_{\alpha}^{x_1}); e \in E\}$ = $F_{P_{\alpha}^{x_1}}$

Noted that soft point F^y and fuzzy point of type P_{α}^{x} with $x \neq y$ we get that

$$f_{F(e)}(z) = \begin{cases} f(z) & if\ z \in F(e) \\ 0 & if\ z \notin F(e) \end{cases}$$

But F(e) = y $\forall x \in X$, so we get f (z) = 0, $\forall z \in F(e) \rightarrow f_{F(e)}(z) = 0$ $\forall z \in X$.

Hence $F_{P_{\alpha}^{x}} = \tilde{\tilde{o}}$.

Example 3.2.6:

Let X = $\{x_1, x_2, x_3, x_4, x_5\}$, F_{x_1}=$\{(e, \{x_1\}), e \in E\}$

For $P_{0.2}^{x_2}$ = $\{(x_1, 0), (x_2, 0.2), (x_3, 0), (x_4, 0), (x_5, 0)\}$

$f_{\{x_1\}}$ = $f_{F(e_1)}$=$f_{F(e_2)}$ = $f_{F(e_3)}$ = $f_{F(e_4)}$ = $\{(e_1, 0), (e_2\ 0), (e_3\ 0), (e_4, 0)\}$

$$f_{\{x_1\}}(y) = \begin{cases} f(y) & if\ y \in \{x_1\} \\ 0 & if\ y \notin \{x_1\} \end{cases} = 0 \qquad \forall x \in X$$

III. The combination of the soft point of type F_e with fuzzy point of type P_{α}^{x} for some $x \in X$, e $\in E$ and $0 < \alpha < 1$, where

$$F_e(y) = \begin{cases} F(e) \neq \emptyset & if\ y = e \\ \emptyset & if\ y \neq e \end{cases}$$

Then we get the following :

1: if $x \in F(e)$ and through it will be a soft fuzzy point of type $F_{P_{\alpha}^{x}}^{e}$

2: if $x \notin F(e)$, then $F_{P_{\alpha}^{x}} = \tilde{\tilde{o}}$

So , through these two conditions we get a point of type

$$e_{F_{P_{\alpha}^{x}}} = \begin{cases} F_{P_{\alpha}^{x}}^{e} & if\ x \in F(e) \\ \tilde{\tilde{o}} & if\ x \notin F(e) \end{cases}$$

Example 3.2.7:

Let $X=\{x_1, x_2, x_3, x_4\}$, $E=\{e_1, e_2, e_3\}$

Let $F_{e_1}=\{(e_1, B), (e_2, \emptyset), (e_3, \emptyset)\}$

$P_{0.2}^{x_1}=\{(x_1, 0.2), (x_2, 0), (x_3, 0), (x_4, 0)\}$

i) If $B=\{x_1, x_3\}$ then

$f_{F(e_1)}=f_{\{x_1, x_3\}}=\{(x_1, 0.2), (x_2, 0), (x_3, 0), (x_4, 0)\}=P_{0.2}^{x_1}$

$f_{F(e_2)}=f_{F(e_3)}=f_{\emptyset}=\tilde{o}$. Thus

$$F_{P_{0.2}^{x_1}}^{e_1}=\{(e_1, P_{0.2}^{x_1}), (e_2, \tilde{o}), (e_3, \tilde{o})\}$$

ii) If $B=\{x_2, x_3, x_4\}$ then

$f_{F(e_1)}=f_B=\{(x_1, 0), (x_2, 0), (x_3, 0), (x_4, 0)\}$

$f_{F(e_2)}=f_{F(e_3)}=f_{\emptyset}=\tilde{o}$.

Thus $F_{P_{0.2}^{x}}=\tilde{\tilde{o}}$

IV. The combination of the soft point type of $F_e^{\,x}$, for some $x \in X$, $e \in E$ with fuzzy point of type p_α, $0<\alpha<1$; Then through it will be a soft fuzzy point of type $F_{P_\alpha^x}^{e}$ because F(e)={x}, so we have $f_{F(e)}(x)=f(x)=\alpha$ and $f_{F(e)}(y)=0 \ \forall\, y \neq x$.

Example 3.2.8 :

Let $X=\{x_1, x_2, x_3, x_4\}$, $E=\{e_1, e_2, e_3\}$ and $P_{0.2}=\{(x_1, 0.2), (x_2, 0.2), (x_3, 0.2), (x_4, 0.2)\}=\{(x, 0.2), \forall\, x \in X\}$

$F_{e_1}^{x_1}=\{(e_1, \{x_1\}), (e_2, \emptyset), (e_3, \emptyset)\}$

$$f_{F_{(e_1)}}(y)=f_{\{x_1\}}(y)=\begin{cases} f(y) & if\ y \in \{x_1\} \\ 0 & if\ y \notin \{x_1\} \end{cases}$$

$f_{F(e_1)}=\{(x_1, 0.2), (x_2, 0), (x_3, 0)\}$

$f_{F(e_2)}=f_{F(e_3)}=f_{\emptyset}=\tilde{o}$

Thus $F_{P_\alpha^{x_1}}^{e_1}=\{(e_1, P_{0.2}^{x_1}), (e_2, \tilde{o}), (e_3, \tilde{o})\}$

if $F_{e_2}^{x_3}=\{(e_1, \emptyset), (e_2, \{x_3\}), (e_3, \emptyset)\}$

$f_{F(e_1)}=f_{\emptyset}=\tilde{o}$

$$f_{F(e_2)} = f_{\{x_3\}} \to f_{\{x_3\}}(y) = \begin{cases} f(x_3) & if\ y = x_3 \\ 0 & if\ y \neq x_3 \end{cases}$$

$f_{F(e_2)} = \{(x_1, 0), (x_2, 0), (x_3, 0.2)\} \to The\ soft\ fuzzy\ point\ \ F^{e_2}_{P^{x_3}_{\alpha}}$

V) The combination of the soft point of type F^x , for some $x \in X$, with fuzzy point of type p_α for $0 < \alpha < 1$. then through it will be a soft fuzzy point of type $F_{P^x_\alpha}$ such that $\forall\ e \in E,\ f_{F(e)} = P^x_\alpha$

Example 3.2.9:

Let X $= \{x_1, x_2, x_3, x_4\}$, E $= \{e_1, e_2, e_3\}$

$$F^{x_1} = \{(e_1, \{x_1\}), (e_2, \{x_1\}), (e_3, \{x_1\})\}$$

$$P_\alpha = \{(x_1, \alpha), (x_2, \alpha), (x_3, \alpha)\}$$

Then the membership function of P_α is

$$f(x) = \alpha \qquad \forall x \in X \quad , \quad 0 < \alpha < 1$$

$$f_{F(e_1)} = f_{\{x_1\}}(y) = \begin{cases} \alpha & if\ y = x_1 \\ 0 & if\ y \neq x_1 \end{cases}$$

$$f_{F(e_1)} = f_{\{x_1\}} = \{(x_1, \alpha), (x_2, 0), (x_3, 0), (x_4, 0)\}$$

$$f_{F(e_2)} = f_{\{x_1\}} = \{(x_1, \alpha), (x_2, 0), (x_3, 0), (x_4, 0)\}$$

$$f_{F(e_3)} = f_{\{x_1\}} = P^{x_1}_\alpha .$$

Then $\qquad f_{F(e_1)} = f_{F(e_2)} = f_{F(e_3)} = P^{x_1}_\alpha$

$$F_{P^{x_1}_\alpha} = \{(e_1, P^{x_1}_\alpha), (e_2, P^{x_1}_\alpha), (e_3, P^{x_1}_\alpha)\}$$

VI) The combination of the soft point of type F_e , for some $e \in E$ with fuzzy point of type p_α ، $0 < \alpha \leq 1$, then it will a soft fuzzy point with a formula $F^e_{P^{F(e)}_\alpha}$.That is

$$F^e_{P^{F(e)}_\alpha}(\epsilon) = \begin{cases} P^{F(e)}_\alpha & if\ \epsilon = e \\ \tilde{o} & if\ \epsilon \neq e \end{cases}$$

where

$$P^{F(e)}_\alpha(x) = \begin{cases} \alpha & if\ x \in F(e) \\ 0 & if\ x \notin F(e) \end{cases}$$

Example 3.2.10:

Let X $= \{x_1, x_2, x_3, x_4\}$, E $= \{e_1, e_2, e_3\}$ and $P_\alpha = \{(x, \alpha) : x \in X\}$

i) If $F_{e_1} = \{(e_1, \{x_1, x_2\}), (e_2, \emptyset), (e_3, \emptyset)\}$

$f_{F(e_1)} = f_{\{x_1,x_2\}} = \{(x_1, \alpha), (x_2, \alpha), (x_3, 0), (x_4, 0)\} = P_\alpha^{F(e_1)}$

$f_{F(e_2)} = f_{F(e_3)} = f_\emptyset = \tilde{o}$.

Thus $F^{e_1}_{P_\alpha^{F(e_1)}} = \{\left(e_1, P_\alpha^{F(e_1)}\right), (e_2, \tilde{o}), (e_3, \tilde{o})\}$

ii) If $F_{e_1} = \{(e_1, \{x_2, x_3\}), (e_2, \emptyset), (e_3, \emptyset)\}$

$f_{F(e_1)} = f_{\{x_2,x_3\}} = \{(x_1, 0), (x_2, \alpha), (x_3, \alpha)\} = P_\alpha^{F(e_1)}$

$f_{F(e_2)} = f_{F(e_3)} = f_\emptyset = \tilde{o}$.

Thus $F^{e_1}_{P_\alpha^{F(e_1)}} = \{\left(e_1, P_\alpha^{F(e_1)}\right), (e_2, \tilde{o}), (e_3, \tilde{o})\}$

VII) The combination of the soft point F_e^x with the fuzzy point P_α^A for $0 < \alpha \leq 1$, we get that:

1- If $x \in A$, then we get the soft fuzzy point of kind $F^e_{P^x_\alpha}$

2- If $x \notin A$, then we get the null soft fuzzy set.

Example 3.2.11: Let

$X = \{x_1, x_2, x_3, x_4\}, E = \{e_1, e_2, e_3\}, F_{e_1}^{x_1} = \{(e_1, \{x_1\}), (e_2\emptyset), (e_3, \emptyset)\}, A = \{x_1, x_2\}, P_{0.3}^A = \{(x_1, 0.3), (x_2, 0.3), (x_3, 0), (x_4, 0)\}$ and

$P_{0.3}^B = \{(x_1, 0), (x_2, 0.3), (x_3, 0), (x_4, 0.3)\}$ with $B = \{x_3, x_4\}$. Then we have

a) The combination of $F_{e_1}^{x_1}$ with $P_{0.3}^A$ we get $F^{e_1}_{P^{x_1}_{0.3}}$

b) The combination of $F_{e_1}^{x_1}$ with $P_{0.3}^B$ we get

$$f_{F(e_1)} = f_{\{x_1\}} = \{(x_1, 0), (x_2, 0), (x_3, 0), (x_4, 0)\}$$

$f_{F(e_2)} = f_{F(e_3)} = f_\emptyset = \tilde{0}$ null fuzzy set.

VIII) The combination of the soft point F^x with the fuzzy point P_α^A for $0 < \alpha \leq 1$, we get that:

1- The soft fuzzy point of kind $F_{P^x_\alpha}$, $x \in A$

2- If $x \notin A$, then the combination is null soft fuzzy set.

Example 3.2.12:

From example 3.2.11, if we take $F^{x_1} = \{(e_1, \{x_1\}), (e_2, \{x_1\}), (e_3, \{x_1\})\}$, the combination with fuzzy point $P^A_{0.3}$ and $P^B_{0.3}$ we have:

a) $f_{F(e_1)} = f_{F(e_2)} = f_{F(e_3)} = \{(x_1, 0.3), (x_2, 0), (x_3, 0), (x_4, 0)\} = P^{x_1}_{0.3}$ because $x_1 \in A$.

b) Since $x_1 \notin B$ we have $f_{F(e_1)} = f_{F(e_2)} = f_{F(e_3)} = \tilde{0}$ null full set

Thus, we get that $F_{P^{x_1}_{0.3}}$ for the combination of $P^A_{0.3}$ with F^{x_1}, and null fuzzy sets for the companion of $P^B_{0.3}$ with F^{x_1}.

IX) : The combination of the soft point F_e with the fuzzy point P^A_α for $0 < \alpha \leq 1$ we get that:

1- If $F(e) \cap A = \emptyset$, then the combination is null soft fuzzy set.

2- $F(e) \cap A \neq \emptyset$, then the combination is soft fuzzy point of kind $F^e_{P_\alpha^{(F(e)\cap A)}}$.

Example 3.2.13:

Let $X = \{x_1, x_2, x_3, x_4\}, E = \{e_1, e_2, e_3\}, A = \{x_1, x_2\}$,

$P^A_{0.2} = \{(x_1, 0.2), (x_2, 0.2), (x_3, 0), (x_4, 0)\}$ and

$F_{e_1} = \{(e_1, \{x_2, x_3\}), (e_2, \emptyset), (e_3, \emptyset)\}$.

a) The combination of $P^A_{0.2}$ and F_{e_1} is $F^{e_1}_{P^{x_2}_{0.2}}$ because $f_{F(e_1)} = f_{\{x_2,x_3\}} =$

$\{(x_1, 0), (x_2, 0.2), (x_3, 0), (x_4, 0)\} = P^{x_2}_{0.2}$

$f_{F(e_2)} = f_{F(e_3)} = f_\emptyset = \tilde{0}$ null fuzzy set.

b) The combination of $P^A_{0.2}$ and F_{e_3} where

$F_{e_3} = \{(e_1, \emptyset), (e_2, \emptyset), (e_3, \{x_2, x_3\})\}$ is null soft fuzzy set because

$f_{F(e_1)} = f_{F(e_2)} = f_\emptyset = \tilde{0}$ null fuzzy set and

$f_{F(e_3)} = f_{\{x_3,x_4\}} = \{(x_1, 0), (x_2, 0), (x_3, 0), (x_4, 0)\}$.

Thus $F(e_1) \cap A = \{x_2\}$. So, the combination is

$F^{e_1}_{P^{x_2}_{0.2}} = F^{e_1}_{P_{0.2}^{(F(e_1)\cap A)}}$, but $F(e_3) \cap A = \emptyset$. So, the combination is null soft fuzzy set.

Table (3-1) The combination of soft points and fuzzy points

Fuzzy points	Soft points	Soft fuzzy points
p_α^x	F_e^x	$F_{p_\alpha^x}^e$
p_α	F_e^x	
p_α^A	$F_e^x, x \in A$	
p_α^x	F^x	$F_{p_\alpha^x}$
p_α	F^x	
p_α^A	$F^x, x \in A$	
p_α^x	F_e	${}^eF_{p_\alpha^x}$
p_α	F_e	$F^e_{p_\alpha^{F(e)}}$
p_α^A	F_e and $F(e) \cap A \neq \emptyset$	$F^e_{p_\alpha^{(A\cap F(e))}}$

Conclusion 3.2.14:

A new five different types of soft fuzzy points via the combination between three different types of soft points F_e^x, F^x *and* F_e with three different types of fuzzy points p_α^x, p_α^A *and* p_α with some definitions and illustrative examples are introduced .

3.3 Soft Fuzzy Topological Spaces

An analytical study for the combination of the soft topological spaces and fuzzy topological spaces is presented with the reformulation of some definitions and using the new soft fuzzy points which presented in section 3.2 to introduce new definition of separation axioms. Also base, subbase for these spaces and soft fuzzy compact space with theorems and illustrative examples

Definition 3.3.1:

A Soft fuzzy topology τ on (X , E) is a family of soft fuzzy sets over (X , E) satisfying the following properties:

1. $\tilde{\tilde{\emptyset}}, \tilde{\tilde{X}} \in {}_s\tau_f$
2. $If\ {}_sA_f,\ {}_sB_f \in {}_s\tau_f\ \ then\ {}_sA_f \cap {}_sB_f \in {}_s\tau_f$
3. $If\ {}_sA_{\alpha f} \in {}_s\tau_f\ \ for\ all\ \alpha \in \Lambda\ ,an\ index\ set,\ then\ \bigcup_{\alpha\in\Lambda} {}_sA_{\alpha f} \in$

   ${}_s\tau_f$. *For any index* Λ and ${}_sA_{\lambda f} \in {}_s\tau_f$

   So that

i. A sub collection ${}_s\beta_f$ of ${}_s\tau_f$ is called base for ${}_s\tau_f$ if for any soft fuzzy open set is union of some members of $s\beta_f$

ii. A sub collection ${}_s\omega_f$ is called sub base for ${}_s\tau_f$ if for any soft fuzzy topology ${}_s\tau_f$ the finite intersections of members of ${}_s\omega_f$ form a basis of ${}_s\tau_f$.

Example 3.3.2 :

Let X ={ x_1, x_2, x_3} and E = {e_1, e_2, e_3}

$F(e_1) = \{x_1\}$

$F(e_2) = \{x_1, x_2\}$

$F(e_3) = \{x_2, x_3\}$ where

F : E → P(x)

$F_A = \{ (e_1, \{x_1\}), (e_2, \emptyset), (e_3, \{x_2, x_3) \}$

$G_A = \{ (e_1, \{x_1, x_2\}), (e_2, \{x_2\}), (e_3, \{x_2\}) \}$

$H_A = \{ (e_1, \{x_1\}), (e_2, \emptyset), (e_3, \{x_2\}) \}$

$J_A = \{ (e_1, \{x_1, x_2\}), (e_2, \{x_2\}), (e_3, \{x_2, x_3\}) \}$

$T_S = \{\bar{\bar{\Phi}}, \bar{\bar{X}}, F_A, G_A, H_A, J_A\}$

$\tilde{A}$={ $(x_1, 0.1), (x_2, 0), (x_3, 0.3)$ } with membership f

$\tilde{B}$={ $(x_1, 0.2), (x_2, 0), (x_3, 0.1)$ } with membership g

$\tilde{C}$={ $(x_1, 0.2)$, $(x_2, 0)$, $(x_3, 0.3)$ }   with membership h

$\tilde{D}$={ $(x_1, 0.1)$, $(x_2, 0)$, $(x_3, 0.1)$ }   with membership k

${}_f\tilde{\mathrm{T}}$={ $\tilde{0}, \tilde{1}, \tilde{A}, \tilde{B}, \tilde{C}, \tilde{D}$}.

Now,

$$F_X \rightarrow \begin{cases} \tilde{A} & {}_fX_A = \{ (e_1, \tilde{A}), (e_2, \tilde{A}), (e_3, \tilde{A})\} \\ \tilde{B} & {}_gX_A = \{ (e_1, \tilde{B}), (e_2, \tilde{B}), (e_3, \tilde{B})\} \\ \tilde{C} & {}_hX_A = \{ (e_1, \tilde{C}), (e_2, \tilde{C}), (e_3, \tilde{C})\} \\ \tilde{D} & {}_kX_A = \{ (e_1, \tilde{D}), (e_2, \tilde{D}), (e_3, \tilde{D})\} \end{cases}$$

$$F_A \rightarrow \begin{cases} \tilde{A} & {}_fF_A = \{ (e_1, P_{0.1}^{x_1}, (e_2, \tilde{0}), (e_3, P_{0.3}^{x_3}) \} \\ \tilde{B} & {}_gF_A = \{ (e_1, P_{0.2}^{x_1}, (e_2, \tilde{0}), (e_3, P_{0.1}^{x_3}) \} \\ \tilde{C} & {}_hF_A = \{ (e_1, P_{0.2}^{x_1}, (e_2, \tilde{0}), (e_3, P_{0.3}^{x_3}) \} \\ \tilde{D} & {}_kF_A = \{ (e_1, P_{0.1}^{x_1}, (e_2, \tilde{0}), (e_3, P_{0.1}^{x_3}) \} \end{cases}$$

$$G_A \rightarrow \begin{cases} \tilde{A} & {}_fG_A = \{ (e_1, P_{0.1}^{x_1}, (e_2, \tilde{0}), (e_3, \tilde{0}) \} \\ \tilde{B} & {}_gG_A = \{ (e_1, P_{0.2}^{x_1}, (e_2, \tilde{0}), (e_3, \tilde{0}) \} \\ \tilde{C} & {}_hG_A = \{ (e_1, P_{0.2}^{x_1}, (e_2, \tilde{0}), (e_3, \tilde{0}) \} \\ \tilde{D} & {}_kG_A = \{ (e_1, P_{0.1}^{x_1}, (e_2, \tilde{0}), (e_3, \tilde{0}) \} \end{cases}$$

$$\mathrm{H_A} \rightarrow \begin{cases} \tilde{A} & {}_fH_A = \{ (e_1, P_{0.1}^{x_1}, (e_2, \tilde{0}), (e_3, \tilde{0}) \} \\ \tilde{B} & {}_gH_A = \{ (e_1, P_{0.2}^{x_1}, (e_2, \tilde{0}), (e_3, \tilde{0}) \} \\ \tilde{C} & {}_hH_A = \{ (e_1, P_{0.2}^{x_1}, (e_2, \tilde{0}), (e_3, \tilde{0}) \} \\ \tilde{D} & {}_kH_A = \{ (e_1, P_{0.1}^{x_1}, (e_2, \tilde{0}), (e_3, \tilde{0}) \} \end{cases}$$

$$\mathrm{J_A} \rightarrow \begin{cases} \tilde{A} & {}_fJ_A = \{ (e_1, P_{0.1}^{x_1}, (e_2, \tilde{0}), (e_3, P_{0.3}^{x_3}) \} \\ \tilde{B} & {}_gJ_A = \{ (e_1, P_{0.2}^{x_1}, (e_2, \tilde{0}), (e_3, P_{0.1}^{x_3}) \} \\ \tilde{C} & {}_hJ_A = \{ (e_1, P_{0.2}^{x_1}, (e_2, \tilde{0}), (e_3, P_{0.3}^{x_3}) \} \\ \tilde{D} & {}_kJ_A = \{ (e_1, P_{0.1}^{x_1}, (e_2, \tilde{0}), (e_3, P_{0.1}^{x_3})\}. \end{cases}$$

That is:

$${}_s\tau_f = \{\tilde{\tilde{\Phi}}, \tilde{\tilde{X}}, {}_fX_A, {}_gX_A, {}_hX_A, {}_kX_A, {}_fF_A, {}_gF_A, {}_hF_A, {}_kF_A, {}_fG_A, {}_gG_A,$$

${}_hG_A$, ${}_kG_A$, ${}_fH_A$, ${}_gH_A$, ${}_hH_A$, ${}_kH_A$, ${}_fJ_A$, ${}_gJ_A$, ${}_hJ_A$, ${}_kJ_A$ }.

Definition 3.3.3 :

A collection ${}_s\beta_f$ of some soft fuzzy sets over (U, E) is called a soft fuzzy open base or simply a base for some soft fuzzy topology on (U, E) if the following conditions hold:

1. $\Phi \in \beta$
2. $\cup \beta = \tilde{E}$ i.e. for each $e \in E$ and $x \in U$, there exists $F_A \in \beta$ such that $\mu^e_{F_A} = 1$
3. If $F_A, G_B \in \beta$ then for each $e \in E$ and $x \in U$, there exists $H_C \in \beta$ such that $H_C \sqsubseteq F_A \sqcap G_B$ and $\mu^e_{F_C} = \min \{\mu^e_{F_A}, \mu^e_{G_B}\}$ where $C \subset A \cap B$.

Note 3.3.4 :

1- From definition.1.3.1 and definition 1.7.1 we conclude that the soft fuzzy topology is extract from combination of soft topology with fuzzy topology.

2- From definition 1.3.3 and definition 1.7.3 we see that the base of soft fuzzy topology is output of the combination of base of soft topology with base of fuzzy topology.

3- From definition 1.3.4 and definition 1.7.4, we see that the sub base of soft fuzzy topology the output of the combination of subase of soft topology with subase of fuzzy topology.

3.4. Soft Fuzzy Separation Axioms

Here various soft fuzzy separation axioms for a soft fuzzy topological spaces with definitions and theorems are introduced.

Note 3.4.1 :

The properties of the separation axioms are defined on the different soft fuzzy points and the definition of soft fuzzy set comes on view of the different soft fuzzy points that is either on two different points of the same type or two points from two different types or valid on any two points from the types and this means that it is more general, that is there are three definitions for T_0 ,T_1,T_2, regular, T_3 , normal and the other kinds of separation axioms.

Theorem 3.4.2:

If the soft fuzzy topology $(\tilde{1}, {}_s\tau_f)$ is ${}_s fT_\circ$-space w.r.t the soft fuzzy point (simply sfp) of the type $F^{e}_{P^{x}_{\alpha}}$ iff the spaces that are combination to configure the fuzzy topology and soft topology be $fT_\circ$- space w.r.t fuzzy point (simply fP) of type P^{x}_{α} and ST_0 – space w.r.t. soft point of type F^{e}_{x} respectively.

Proof. Since, for any two different soft fuzzy points of the type $F^{e}_{P^{x}_{\alpha}}$ are coming from the combination of

i) $P^{x_1}_{\alpha} \neq P^{x_2}_{\alpha}$ *and* $F^{e_1}_{x_1} \neq F^{e_1}_{x_2}$

ii) $P^{x_1}_{\alpha} \neq P^{x_2}_{\lambda}$ *and* $F^{e_1}_{x_1} \neq F^{e_2}_{x_2}$

iii) $P^{x_1}_{\alpha} \neq P^{x_1}_{\lambda}$ *and* $F^{e}_{x_1}$

iv) $P^{x_2}_{\alpha}$ *and* $F^{e_1}_{x_2} \neq F^{e_2}_{x_2}$

Assume that $(\tilde{\tilde{X}}, \tilde{\tilde{\tau}})$ is ${}_s fT_\circ$-space , then for any two distinct fuzzy points

$a) P^{x_1}_{\alpha} \neq P^{x_2}_{\alpha}$

$b) P^{x_1}_{\alpha} \neq P^{x_2}_{\lambda}$

$c) P^{x_1}_{\alpha} \neq P^{x_1}_{\lambda}$

And from i,ii.iii,iv, above we get the following distinct soft fuzzy pinots:

1) $F^{e_1}_{P^{x_1}_{\alpha}} \neq F^{e_1}_{P^{x_1}_{\alpha}}$

2) $F^{e_1}_{P^{x_1}_{\alpha}} \neq F^{e_2}_{P^{x_2}_{\lambda}}$

3) $F^{e_1}_{P^{x_1}_{\alpha}} \neq F^{e_2}_{P^{x_1}_{\lambda}}$

But $(\tilde{\tilde{\tau}}, {}_s T_f)$ is a s $fT_\circ - space$

∴ ∃ a Sf-open set ${}_sG_f$ contains one of the two points and does not contains the other, but ${}_sG_f$ is obtained from the combination of $\tilde{G}$ f-open with G_A s-open so $\tilde{G}$ f-open contains one of the two fuzzy points in (a,b,c) and does not contains the other and this gives that $(\tilde{1}, {}_f\tilde{\tau})$is fTo-space and by the same method , we get $(\tilde{X}, {}_s\tilde{\tau})$is too ${}_sT_\circ$-space

⇐ suppose that $(\tilde{X}, {}_s\tilde{\tau})and(\tilde{1}, {}_f\tilde{\tau})$ are sT_o-space and fT_o-space respectively.

Now, let $F^{e_1}_{P^{x}_{\alpha}} \neq F^{e_2}_{P^{\alpha}_{x}}$ and (1),(2),(3) be different soft fuzzy points that can be get by the combination of the soft points with the fuzzy points, but $(\tilde{1}, {}_f\tilde{\tau})$,

$(\tilde{X},\ _S\tilde{\tau})$,so $\exists$ a f- open set $\tilde{B}$ contains one of the two fuzzy points and does not contains the another and s-open set $F_A$contains one of the two soft points and does not contains the another , By the combination between $\tilde{B}$ with F_A we obtain sf-open set $_S\tilde{B}_{F_{Af}}$ which contains one of the two soft fuzzy points and does not contains the another .

Note 3.4.3: The theorem is valid for all types of soft fuzzy points mentioned above.

Theorem 3.4.4:
The soft fuzzy topology $_S\tau_f$ is $_SfT_1$-space w.r.t the s.f.p. of the type $F^e_{P^x_\alpha}$ iff the spaces that are combination to configure the fuzzy topology and of type soft topology be fT_1- space w.r.t. f.p. of type P^x_α and ST_1 – space w.r.t sp of type F^e_x iff for two distinct soft fuzzy points sets, there exists two soft fuzzy open each one contains only one soft fuzzy point and does not contain the other.
Proof. It is similar to the proof of Theorem 3.4.2.

Theorem 3.4.5:
The soft fuzzy topology $_S\tau_f$ is SfT_2-space (Hausdoorf space) w.r.t. the s.f.p. of the type $F^e_{P^x_\alpha}$ iff the spaces that are combination to configure the fuzzy topology and of type soft topology be fT_2- space w.r.t f.p. of type P^x_α and ST_2- space w.r.t s.p. of type F^e_x iff for two distinct soft fuzzy points there exist two soft fuzzy open sets each one contains only one soft fuzzy point and does not contain the other.

Theorem 3.4.6:
The soft fuzzy topology $_S\tau_f$ is SF Regular space w.r.t the s.f.p. of the type $F^e_{P^x_\alpha}$ iff the spaces that are combination to configure the fuzzy topology and of type soft topology be f- regular space w.r.t. f.p. of type P^x_α and S- regular space w.r.t. s.p. of type F^e_x .

Note 3.4.7: The soft fuzzy topology $(\tilde{\tilde{X}},\ _S\tau_f\ )$ is called SfT_3 if it is soft fuzzy regular space and soft fuzzy $T_1 - space.$

Theorem 3.4.8:

Soft fuzzy topological space is soft fuzzy compact iff it is soft compact and fuzzy compact

Proof :$\Leftarrow$ For any cover of soft fuzzy open sets to the space $\tilde{\tilde{X}}$we get that $soft\ space\ \tilde{X}$ is covered by soft open sets and ${}_f\tilde{1}$ is covered by fuzzy open sets. But the spaces are soft compact and fuzzy compact respectively , so there exists finite sub cover by soft set and fuzzy set , and we may put $m \in N$ which is the number of the soft open sets and $n \in N$ is the number of the fuzzy open sets, so the combination (m,n) is also finite so that (m,n) are sub covers of soft fuzzy open sets for the space $\tilde{\tilde{X}}$. Therefore the space$\tilde{\tilde{X}}$ is soft fuzzy compact .

$\Rightarrow$ Suppose that the space $\tilde{\tilde{X}}$ is soft fuzzy compact .To show that the space $\tilde{\tilde{X}}$ is soft compact and the space ${}_f\tilde{1}$ fuzzy compact, for every soft open cover for the soft space $\tilde{X}$ and every fuzzy open cover to fuzzy space ${}_f\tilde{1}$ then by combination of soft space and fuzzy space we get soft fuzzy space covered by soft fuzzy open sets , so there exists , finite sub cover and again we used the combination to have finite soft cover and finite fuzzy cover respectively. Therefore the soft space is soft compact and fuzzy space is also fuzzy compact.

Proposition 3.4.9:

Let ${}_SA_f$ be a soft fuzzy closed set in soft fuzzy compact space $\left(\tilde{\tilde{X}},\ {}_S\tau_f\ \right)$. Then ${}_SA_f$ is also soft fuzzy compact.

Proof. Let $\mathcal{L}$ be a collection of soft fuzzy open cover to ${}_fF_A$, s the cover $\mathcal{L}$ with complement of ${}_SF_{Af}$ its cover to $\tilde{\tilde{X}}$, by soft fuzzy compact of $\tilde{\tilde{X}}$, there is finite subcover to $\tilde{\tilde{X}}$ which also cover to ${}_SA_f$

Definition 3.4.10:

A soft fuzzy set $F^e_{p^x}(\varepsilon) = \begin{cases} p^x_\alpha\ if\ \varepsilon = e\ for\ 0 < \alpha \le 1 \\ \tilde{0},\ if\ \varepsilon \ne e \end{cases}$

i. A soft fuzzy set $F_{P^x_\alpha} = \{(e, P^x_\alpha), \forall e \in E, 0 < \alpha \le 1\}$ is a soft fuzzy set of type ii

ii. A soft fuzzy set $F_{P_{\alpha}^{x}} = \{(e, P_{\alpha}^{x}), \forall e \in E, 0 < \alpha \leq 1\}$ is a soft fuzzy set of type ii

iii. A soft fuzzy set $F_{P_{\alpha}^{x}}^{e}$ is a soft fuzzy point of type ii

such that $^{e}F_{p_{\alpha}^{x}} = \begin{cases} F_{P_{\alpha}^{x}}^{e} & if\ x \epsilon F_{(e)} \\ \tilde{0} & if\ x \epsilon F_{(e)} \end{cases}$

iv. A soft fuzzy set $F_{p_{\alpha F_{(e)}}^{x}}^{e}$ is a soft fuzzy point of type iv where

$$F_{p_{\alpha}^{F(e)}}^{e}(\varepsilon) = \begin{cases} p_{\alpha}^{F(e)} & if\ \varepsilon = e \\ \tilde{0} & if\ \varepsilon \neq e \end{cases}$$

v. A soft fuzzy set $F_{p_{\alpha}^{(F(e) \cap A)}}^{e}$ is a soft fuzzy point of type **v** such that $F(e) \cap A = \emptyset$.As shown in table (3-1).

Exercises

1- If $F_A \cap G_B \neq \tilde{\varphi}$ and $\tilde{A} \wedge \tilde{B} = \tilde{0}$ .Do you think will be the result? Of ${}_{s}A_{f}$ $\wedge$ ${}_{s}B_{f}$, enhance your answer with an example .

2- Let X = { x , y , z , zz } , E = { e , a , f , ff } , A = { a , f , ff } and
F_A = { (e ,∅) , (a , { x}) , (f , {y,z }) , (ff , { y,zz}) }
$\tilde{A} = \{ (x, 0.1), (y, 0), (z, 0.8), (zz, 0.9)\}$
Find the soft fuzzy set ${}_{s}A_{f}$.

3- Find the soft fuzzy points that combination of the soft and fuzzy points
a- $P_z^{0.4}$ $with$ F_e^y
b- $P_{0.2}$ with F_x
c- $P_{0.5}^{\{x,y\}}$ with F_e
Where X = { x , y , z } , E = { e , a , f } and F (e) = { x , y }

4- Define the interior , exterior , boundary , accumulation and adherent soft fuzzy points as the standard definitions in general topology except this defined in soft fuzzy topological spaces with respect to different soft fuzzy points .

5- Given the relationship between the soft (int. , ext. , br. , limit and adh.) points , fuzzy (int. , ext. br. , limit , adh.) points with soft fuzzy (int. , ext. , br. , limit , adh.) points .

6- Define the neighborhoods via the soft fuzzy topological space with respect to different soft fuzzy points and promote it by some examples .

7- Let ${}_sA_f$ be any soft fuzzy set , so we can define the soft fuzzy closure of it by $cl(\ {}_sA_f\ ) = \cap \{\ {}_sF_f\ ;\ {}_sA_f\ \leq\ {}_sF_f\ ,\ {}_sF_f^C\ \in\ {}_s\tau_f\ \}$. Show that the following properties are hold for any soft fuzzy sets ${}_sA_f,\ {}_sB_f$

   a- If ${}_sA_f\ \leq\ {}_sB_f\ ,then\ cl(\ {}_sA_f) \leq cl(\ {}_sB_f\ )$
   b- $cl(\ {}_sA_f\ ) \vee cl(\ {}_sB_f) = cl(\ {}_sA_f\ \vee\ {}_sB_f)$
   c- $cl(cl(\ {}_sA_f\ )) = cl\ (\ {}_sA_f)$
   d- $cl(\ {}_sA_f\ )$ is soft fuzzy set .

8- For any soft fuzzy open set ${}_sG_f$ and soft fuzzy set ${}_sA_f$, show that $cl(\ {}_sA_f\ ) \wedge\ {}_sG_f \leq cl(\ {}_sA_f \wedge\ {}_sG_f\ )$

9- If we defined the concept of soft fuzzy interior of any soft fuzzy set ${}_sA_f$ by $int(\ {}_sA_f\ ) = \vee\{\ {}_sG_f\ \in\ {}_s\tau_f\ ;\ \ {}_sG_f \leq\ {}_sA_f\ \}$ prove the following properties

   a- $int(int(\ {}_sA_f)) = int(\ {}_sA_f\ )$
   b- ${}_sG_f$ is soft fuzzy open set iff $int(\ {}_sG_f\ ) =\ {}_sG_f$
   c- If ${}_sA_f\ \leq\ {}_sB_f\ ,then\ int\ (\ {}_sA_f\ ) \leq int(\ {}_sB_f)$
   d- $int(\ {}_sA_f\ \wedge\ {}_sB_f\ ) = int(\ {}_sA_f) \wedge int(\ {}_sB_f)$

10-Given any soft fuzzy topology space it explain the soft fuzzy interior and closure by7 some examples .

Chapter Four

Soft Probability and Soft Fuzzy

Reliability in Topological Space

4. Preliminaries

This chapter devoted to study fuzzy probability, Probabilistic interpretation of fuzzy sets, probabilistic soft sets, soft probability and the relationship between fuzzy reliability and classical reliability. In addition to that, soft fuzzy reliability in topological spaces is presented in view of soft approximation and by considering reliability as probability according to its definition and consider the soft fuzzy sets as soft sets in the result according to our belief in current study .

4.1 Fuzzy probability

Probability theory does not give precise answers but only probabilities. The imprecision in probability theory comes from our incomplete data of the system but the random variable (measurements) still have precise value. For example, when we flip a coin we have only a partial knowledge about the physical structure of the coin and the initial conditions of the flip. If our knowledge about the coin were complete , we could predict exactly whether the coin lands heads or tails. However, we still suppose that after the coin is flipped, we can tell precisely whether it is head or tail. In fuzzy probability theory, we also have an imprecision in our measurements, and random variables must be replaced by fuzzy random variables and events by fuzzy events.

Since fuzzy events are essentially fuzzy sets, we start with a comparison of crisp sets and fuzzy sets. In particular, we define complements, intersections, union and differences for fuzzy sets as they were cleared in chapter one .

Let $X = \{x_1, \cdots, x_n\}$ be a finite set and let $pr.$ be a probability function defined on all subsets of X with $pr.(\{x_i\}) = a_i, 1 \leq i \leq n, 0 < a_i < 1$, all i, and $\sum_{i=1}^{n} a_i = 1$. We will substitute a fuzzy number $\bar{a}_i$ for a_i, for some i, to obtain a discrete fuzzy probability distribution. Where do these fuzzy numbers come from?

In some problems, because of the way the problem is stated, the values of all the a_i are crisp and known. For example, consider tossing a fair coin and $a_1 =$ the probability of getting a "head" and $a_2 =$ is the probability of obtaining a "tail". Since we assumed it to be a fair coin we must have $a_1 = a_2 = 0.5$. In this case we would not substitute a fuzzy number for a_1 or a_2. But in many other problems the a_i are not known exactly and they are either estimated from a random sample or they are obtained from "expert opinion".

Suppose we have the results of a random sample to estimate the value of a_1. We would construct a set of confidence intervals for a_1.

Assume that we do not know the values of the a_i and we do not have any data to estimate their values. Then we may obtain numbers for the a_i from some group of experts. This group could consist of only one expert.

First assume we have only one expert and he is to estimate the value of some probability pr. Let $a =$ the "pessimistic" value of pr, or the smallest possible value, let $c =$ the "optimistic" value of pr. We then ask the expert to give values for a, b, c and we construct the triangular fuzzy number $\bar{p} = (a, b, c)$ for pr. If we have a group of N experts all to estimate the value of p we select the a_i, b_i and $c_i, 1 \leq i \leq N$, from them. Let a be the average of the a_i, b is the mean of the b_i and c is the average of the c_i. The simplest thing to do is to use (a, b, c) for pr.

4.1.1 Probabilistic Interpretation of Fuzzy Sets

By viewing membership degree of an element x in a fuzzy set A , it is possible to give A a probabilistic interpretation as the conditional probability of A given x

That is : $\mu_A(x) = \Pr(A|x)$ ......(4-1)

Such interpretation is consistent with the complement operation on fuzzy sets since

$\mu_{A \cup B} = \Pr(A \cup B \mid x)$ ....(4-2)

where $\Pr(A \cup B \mid x) = \Pr(A \mid x) + \Pr(B|x) - \Pr(A \cap B \mid x)$....(4-3)

Hence , it should adopt the probabilistic interpretation , we get

$\mu_{A \cup B}(x) = \mu_A(x) + \mu_B(x) - \mu_{A \cap B}(x)$........(4-5)

This only establishes constraint between the two operations (union ,intersection)

If A and B are conditionally independent of x , that is

$\Pr(A \mid x, B) = \Pr(A| x)$, $\Pr(B| x, A)$ ....(4-6)

$= \Pr(B|x)$ we obtain $\Pr(A \cap B \mid x) = \Pr(A \mid x) \times \Pr(B \mid x)$.(4-7)

with the remark

$$\Pr(A \cap B \mid x) = \begin{cases} \Pr(A \mid x) & \text{if } A \subseteq B \\ \Pr(B \mid x) & \text{if } B \subseteq A \end{cases} \quad \ldots\ldots\ldots(4\text{-}8)$$

Or equivalently

$$\Pr(A \cap B \mid x) = \min\{\Pr(A \mid x), \Pr(B \mid x)\} \ldots\ldots(4\text{-}9)$$

From the case above , we notice that the probability $\Pr(A \cap B \mid x)$ depend not on $\Pr(A \mid x)$, $\Pr(B \mid x)$only but also the relationship between A and B.

So it is impossible to find a general intersection operator

$t \ni \Pr(A \cap B \mid x) = t(\Pr(A \mid x), \Pr(B \mid x)) \ldots\ldots.(4\text{-}10)$

from above we conclude that a major limitation of the probabilistic interpretation of fuzzy sets.

4.2 Probabilistic Soft Sets

In this section the concept of probabilistic soft sets and several illustrative examples are presented. We begin with some notations in probability theory.

Defintinition4.2.1:

Let U be a finite set. A function μ from U to the closed unit interval [0,1] is called a probability distribution on U if $\sum_{x \in U} \mu(x) = 1$.

The set $\{ x \in U : \mu(x) > 0\}$is called the support of μ. For any $x \in U$, we use $\tilde{x}$ to denote the unique probability distribution with $\tilde{x}(x) = 1$. By *D*(U)we denote the set of all probability distributions on the set U. If μ is a probability distribution with support$\{x_1, x_2, \ldots, x_n\}$, we sometimes write μ as $\mu = \frac{\mu(x_1)}{x_1} + \frac{\mu(x_2)}{x_2} + \cdots + \frac{\mu(x_n)}{x_n}$

With this notation, $\tilde{x} = \frac{1}{x}$. We merge soft set theory with probability theory in the following definition.

Definition 4.2.2:

Let U be a universal and E the set of parameters. A probabilistic soft set over U is a pair (F, A) consisting of a subset A of E and a mapping $F: A \to D(U)$

Similar to the standard soft sets, a probabilistic soft set over U is a parameterized family of probability distributions over the set U. The definition above can be readily extended to the cases of general discrete probability distributions and continuous probability distributions by introducing the standard notions of σ-algebra and measurable sets.

For computer storing of a probabilistic soft set in a computer, it may represent a probabilistic soft set by a data table containing rows labeled by parameters of interest, columns labeled by objects in U, and entries of the table indicating the probability that an object occurs under some parameter. That is, if the parameter set E=$\{e_1, e_2, \ldots\}$and U = $\{u_1, u_2, \ldots\}$, then a data table(4-1) is of the following form :

table(4-1) data table

$E\backslash U$	u_1	u_2	
e_1	$F(e_1)(u_1)$	$F(e_1)(u_2)$	$\cdots$
e_2	$F(e_2)(u_1)$	$F(e_2)(u_2)$	$\cdots$
$\vdots$	$\vdots$	$\vdots$	$\vdots$

Noted that the approximate function $F(e)$ is a set called e element of the soft set A .

such that $F(e) = \begin{cases} \emptyset & if\ e \notin A \\ \neq \emptyset & if\ e \in A \end{cases}$

In the above table if $F(e_i) = \emptyset$for some i , then we delete the row that e_i lies in. As a result, we see that the sum of entries in each row is exactly equal to 1; this follows from the property of probability distribution.

Example 4.2.3:

Let us consider the weather forecast for some day. Assume that U={w_1, w_2, . . . , w_8},where w_1="sunny", w_2="sunny interval", w_3="partly cloudy", w_4="mostly cloudy" , w_5="overcast", w_6="light rain", w_7="thunderstorms", and w_8="showers". Let E={morning, afternoon, evening, midnight}. The weather forecast may be represented by a probabilistic soft set (F,A)like this:

A={morning, afternoon, evening, midnight};

F(morning) = $\frac{0.9}{w_1}+\frac{0.1}{w_2}$,

F(afternoon)= $\frac{0.6}{w_4}+\frac{0.3}{w_5}+\frac{0.1}{w_7}$,

F(evening) = $\frac{0.6}{w_5}+\frac{0.4}{w_5}$,

F(midnight) = $\frac{0.2}{w_6}+\frac{0.8}{w_7}$.

The tabular representation of this probabilistic soft set is given in Table(4.2).

Table (4-2): Tabular representation of the probabilistic soft set in example 4.2.3.

$E\backslash U$	w_1	w_2	w_3	w_4	w_5	w_6	w_7	w_8
morning	0.9	0.1	0	0	0	0	0	0
afternoon	0	0	0	0.6	0.3	0	0.1	0
evening	0	0	0	0	0.6	0	0.4	0
midnight	0	0	0	0	0	0.2	0.8	0

Definition 4.2.4:

Let $F_A$and G_B be probabilistic soft sets over a common universal U. We say that F_A is a probabilistic soft subset of G_B, denoted by $F_A\subseteq G_B$ if A⊆B and F(e) ⊆ G(e) for all e ∈ A . The two probabilistic soft sets are called equal, written $F_A= G_B$, if both $F_A\subseteq G_B$ and $G_B\subseteq F_A$.

In probability theory , a probability space (Ω, M, p) which models stochastic process ,where Ω represent a sample space , M represent a collection of events and Pr is the probability measure (M is a σ $algebra$ $over$ Ω)

the probability theory cannot applied successfully since the events in daily experiences are related with several parameters for this problem then it can be considered as a soft sets .
we need to generalized probability theory to soft probability theory[155], this oblique to define the soft events and its probability and defining soft measurable space with its properties and a soft σ –algebra. [105]
because null soft set is not unique and it depends on $A \subseteq E$ (we use $\emptyset_A$)

Definition 4.2.5 :
A collection M of soft sets over the initial universal set X is called a soft σ – algebra on X if M has the properties below :

i. $\widetilde{X} \in M$
ii. if $(F, E) \in M$ then $(F, E)^c \in M$
iii. it is closed under countable union .

note4.2.6 :

i. If M is a soft σ – algebra in X ,then (X, M, E) is called a soft measurable space, and the members of M are called the soft measurable sets in X .
ii. If (X, M, E) is a soft measurable space and (Y, τ, E) is a soft topological space , f is a mapping such that $f: (X, M, E) \rightarrow (Y, \tau, E)$ then f is measurable if for every open set (G, B) in $(y, \tau, E), f^{-1}(G, B)$ is a measurable set in (X, M, E) .

Example 4.2.7 :
Suppose $X = \{x_1, x_2, x_3\}$ and the set of parameter $E = \{e_1, e_2\}$ then $M = \{\tilde{X}, \widetilde{\emptyset}, (F_{1E}), (F_{2E}), (F_{3E}), (F_{4E}), (F_{5E}), (F_{6E})\}$
Is a soft σ – algebra over X where each (F_i, E) is a soft set which is described as follows:

$F_{1E} = \{(e_1, \{x_1\}, (e_2, \emptyset)\}$
$F_{2E} = \{(e_1, \{x_1, x_3\}, (e_2, \{x_2\})\}$
$F_{3E} = \{(e_1, \{x_1, x_3\}, (e_2, X)\}$
$F_{4E} = \{(e_1, \{x_1, x_2\}, (e_2, \{x_1, x_3\})\}$
$F_{5E} = \{(e_1, \{ x_2\}, (e_2, \{x_1, x_3\})\}$
$F_{6E} = \{(e_1, \{ x_3\}, (e_2, \{x_2\})\}$

4.3 Operations with Probabilistic Soft Sets

In this section, we pay attention to some operations with probabilistic soft sets. Unlike the standard soft sets which inherit some set-theoretic properties ,probabilistic soft sets have some probability-theoretic properties.

For later need, let us introduce a notation. For any λ∈[0,1] and $\mu \in D(U)$, define a scalar multiplication $\lambda \cdot \mu: U \longrightarrow [0,1]$ by $(\lambda \cdot \mu)(x) = \lambda \cdot \mu(x)$ for all $x \in U$, where the dot "·" in $\lambda \cdot \mu(x)$ stands for the product of λ and $\mu(x)$. Clearly, the scalar multiplication λμ is not necessarily a probability distribution on U. Similarly, we have the notation μλ, which is defined by $\mu^{\lambda}(x) = (\mu(x))^{\lambda}$ for all $x \in U$. There is many methods for combining probability distributions. For more detail see[148].
We redefine the following definitions which are presented in .

Definition 4.3.1:
Suppose that $(F_i, A), i = 1,2,\ldots,n$, is a probabilistic soft set over a universal U and F_i (e)$\neq \emptyset$ for all e ∈A. Given w_i∈[0,1], $i = 1,2,\ldots,n$, with $\sum_{i=1}^{n} w_i = 1$, the additive averaging of (F_i,A) (with respect to w_i), $i = 1,2,\ldots,n$, is a probabilistic soft set (F, A) over U, where F:$A \longrightarrow D(U)$is defined by
F(e) = $\sum_{i=1}^{n} w_i F_i$ (e) (4-11)
for every e ∈A.
Intuitively, it is regard *n* as the number of experts, and view F_i (e)as expert *i's* probability distribution for the uncertain quantity e, where expert *i* has the weight w_i. The weights in (4-11) clearly can be used to represent, in some sense, the relative quality of the different experts, or sources of information. In the simplest case, the experts are viewed as equivalent ,and (4-11) becomes a simple arithmetic average. If some experts are viewed as "better" than others (in the sense of being more precise due to having better information, for example), the "better" experts might be given greater weights. The determination of the weights is a subjective matter, and numerous interpretations can be given to the weights.

Definition 4.3.2:

Suppose that $(F_i, A), i = 1,2,\ldots,n,$ is a probabilistic soft set over a universal U and $F_i(e) \neq \emptyset$ for all e ∈ A. The multiplicative averaging of $(F_i, A), i = 1,2,\ldots,n,$ is a probabilistic soft set (F, A) over $U,$ where $F: A \dashrightarrow D(U)$ is defined by

$$F(e) = k \prod_{i=1}^{n} Fi(e)^{wi} \quad \ldots\ldots\ldots\ldots\ldots\ldots \quad (4\text{-}12)$$

For every e ∈A. In (4-12), k is a constant and the weights w_i satisfy some restrictions to assure that F(e) is a probability distribution over U. Typically, the weights w_i are restricted to summation to 1.

A new operation with probabilistic soft sets, called conditional probabilistic soft sets. Recall that in probability theory, conditional probability (measure) is a wonderful tool to update probability measures when some new information is taken into account. Formally, the probability of V conditional on W, denoted by $\mu W(V)$, is defined as

$$\mu W(V) = \frac{\mu(V \cap W)}{\mu(W)} \quad \ldots\ldots\ldots (4-13)$$

$when \;\; \mu(W) > 0.$

Definition 4.3.3:

Suppose that F_{iA} is a probabilistic soft set over a universal U and W⊆U. The conditional probabilistic soft set (F_W, A)is a probabilistic soft set over U, where F_W:A→ D(U) is defined by

$$F_W(\mathrm{e}) = \begin{cases} F_W(\mathrm{e}), & \text{if } F_W(\mathrm{e}) > 0 \\ F_E(\mathrm{e}), & \text{otherwise.} \end{cases} \quad \ldots\ldots\ldots (4-14)$$

4.4 Fuzzy Reliability

Definition 4.4.1:

Fuzzy reliability is the ability of a device performing its purpose in varying degree of success for the period of time intended under the operating conditions encountered.

The elements ability, device , performance in varying degrees of success , time , and operating conditions . This performance is called (fuzzy performance).

In view of the fuzzy probability theory, the problems of fuzzy reliability are categorized to three modes as follows.

i. Fuzzy event – accurate probability
ii. Clear event – fuzzy probability
iii. Fuzzy event – fuzzy probability

Notes from definition above4.4.2:

i. There is no mistakes what the fuzzy reliability is purposed at.
ii. The fuzzy reliability is development of the classical reliability.
iii. The fuzzy performances may be resembled by fuzzy subsets by means of the construction of fuzzy language rule in fuzzy mathematics since it is fuzzy performances are represented by (large) and (small) by using the language operators to distinguish the varying degree of large and small performance. For more detail see[152]
iv. Applying the concept of complement set and operation rules of the language value in order to more differ the fuzzy performance subsets. i.e the different fuzzy performances may be represented by different fuzzy subsets.

In terms of the definition of fuzzy conditional probability and according to the definitions of classical reliability R and fuzzy reliability $\tilde{R}$ represented, we have:

pr.(A $\wedge\tilde{A}_i$) $= pr.(\tilde{A}i|A)$ pr.(A)...... (4-15)

Where $\tilde{A}$ is one of a fuzzy performance subsets and

pr.(A)=R , $Pr.(A \wedge \ \tilde{A}_i$)=$\tilde{R}$ ....(4-16)

from equations. (4-15) and (4-16) ,we get

$\tilde{R} = Pr.\left(\tilde{A}_i|A\right)R \ldots\ldots\ldots\ldots\ldots\ldots\ldots..(4-17)$

Let $\mu_{\tilde{A}_i}(R)$ is the degree of membership of R in $\tilde{A}i$ and by substituting $\mu_{\tilde{A}i}(R)$ for pr. $\left(\tilde{A}_i|A\right)$ finally we get the equation

$\tilde{R} = \mu_{\tilde{A}_i}(R)$ R ....(4-18)

Which represents the relation between classical reliability R and fuzzy reliability $\tilde{R}$.

4.5 Soft Fuzzy Reliability in Topological Space

In this section we study stochastic approximation spaces from topological view. Since the approximation space $S = (U, R\ )$ with relation R defines a unique topology $_S\tau$, where U/R is a sub base of τ_S, then the order triple $K = (U, R, pr)$ is called the stochastic approximation space, where R is a relation and pr is a probability measure ,we study soft fuzzy reliability in topological space in the view of the soft set theory and the definition of soft fuzzy sets with the properties of soft fuzzy sets which presented in chapter three in detail .

Soft Approximation4.5.1

Soft set theory has potential applications in many different fields like probability theory. A soft set is replaced instead of an equivalence relation and it can be used to granulate the universal as shown in definition below.

Definition 4.5.1.1:

Let (U ,R) be a pawlak approximation apace and (F,A) be a soft set over U. The lower and upper rough approximation of (F,A) with respect to (U,R) are denoted by $R_*(\epsilon)$= F_{*A} and $R^*(\epsilon) = F_A^*$ which are soft sets over U with the set- valued mapping given by :

$F_*\ (x) = R_*(F(x))$ and $F^*(x)=\ \ R^*(F(x))$ where $x \in A$, R_* is lower approximation and R^* is upper approximation .

There is a natural question , Is there any link between soft set theory and other theories?

In Molodtsov definition . of soft set any subset of U $\times$ U is called a binary relation on U. if R is a binary relation on U , since R is reflexive, symmetric and transitive , then R is called an equivalence relation The definition below of approximation space is in Pawlak's sense gives the answer .

Definition 4.5.1.2:

Let U be an universal set and let R be an equivalence relation on U called indiscernibility relation. The pair (U , R) is called soft approximation space.

Definition 4.5.1.3:

Let $F_A$be a soft set over$U \times U$, then F_A is called a soft binary relation over U.

Definition 4.5.1.4:

A soft binary relation F_A over a set U is called a soft equivalence relation over U if $F(\propto_i) \neq \emptyset$ is an equivalence relation over U $\forall \propto_i \in A$. Every equivalence relation on a set partitions its into disjoint classes and any partition of the set supply us on equivalence on the set.

Therefore a soft equivalence relation over U supply us a parameterized collection of partitions of U as in the following illustrative example .

Example 4.5.1.5:

Let X = $\{h_1, h_2, h_3, h_4, h_5, h_6\}$ be a set of houses , E is the set of parameters

Let X = $\{h_1, h_2, h_3, h_4, h_5, h_6\}$ be a set of houses , E is the set of parameters
for the selection of the house

E={beautiful(e_1), expensive (e_2) , in the green surroundings(e_3) , modern(e_4), cheap(e_5)} and A= $\{e_1, e_2, e_3, e_4\}$, A $\subseteq$ E

Let F_A be the soft set to classify the houses with respect to parameters given by A , such that :

F(e_1) = $\{h_1, h_2, h_3\}$, F(e_2) =$\{h_1, h_3\}$,

F(e_3) = $\{h_1, h_3, h_4, h_5\}$, F(e_4) =$\{h_1, h_3, h_6\}$

For computer applications the soft set(F,A)is represented in the following table which represent the tabular representation of soft sets (Lin et al tabular)

Table (4-3) the tabular representation

$E\backslash U$	h_1	h_2	h_3	h_4	h_5	h_6
e_1	1	1	1	0	0	0
e_2	1	0	1	0	0	0
e_3	1	0	1	1	1	0
e_4	1	0	1	0	0	1

From the table above each of the parameters $e_1, e_2, e_3, and\ e_4$ induces an equivalence relation on X. denote it by (R , A) . An equivalence classes were obtained for each of the equivalence relations as in table below :

Table (4-4) Equivalence relation and obtained equivalence classes

Equivalence relation	Equivalence classes
$R(e_1)$	$\{h_1, h_2, h_3\}$, $\{h_4, h_5, h_6\}$
$R(e_2)$	$\{h_1,\ h_3\}$, $\{h_2, h_4, h_5, h_6\}$
$R(e_3)$	$\{h_1, h_3, h_4, h_5\}$, $\{h_2,\ h_6\}$
$R(e_4)$	$\{h_1, h_3, h_5\}$, $\{h_2, h_4, h_6\}$

Observe that there is an indiscernibility (briefly IND) relation defined by the soft set F_A it self. This indiscernibility relation is got by the intersection of all the equivalence relations induced by parameters , we can write

INDF_A= $\bigcap_{e_{i\in A}} R(e_i) = R$

Hence the partition of X obtained by

IND F_A is $\{h_1,\ h_3\}$, $\{h_2\}$, $\{h_4,\ h_5\}$, $\{h_6\}$

Notice that for any R(e_i) , i=1,2,3,4 (X,R(e_i)) gives an approximation spaces . So that (X , R) from example above any subset of X can be approximated by the equivalence relation R($\propto_i$) . The equivalence class containing an element $x\in X$ determined by the equivalence relation on R($\propto_i$) is denoted by $[x]_{R(\propto_i)}$. The parameterized collection of subsets denoted by (R^x, A) defined as $R^x =$ ($\propto_i$)=$\cup\{[x]_{R(\propto_i)} : [x]_{R(\propto_i)} \subseteq X\}, \forall_{\propto_i \in A}$ is called soft lower approximation of x with respect to soft equivalence relations The parameterized collection , ($\tilde{R}^x$,A) of subsets of X denoted by

($\tilde{R}^x$,A)and defined by $\tilde{R}^x(\propto_i)$= $\cup\{[x]_{R(\propto_i)} \cap X \neq \emptyset\}\forall_{\propto_i \in A}$ is called soft upper approximation of X with respect to soft equivalence relations (R,A).According to the above sense , we denote $(\underline{R}^x, A) = R_*(A), (\bar{R}^x, A) = R^*(A)$,

$Pr.R_*(A) = \underline{Pr}\ (A) = Pr\ [int\ (A)] and\ Pr.R^*(A) = \overline{Pr}\ (A) = Pr\ [cl(A)]$,

the soft probability $\widetilde{\widetilde{pr}}$ of A is $(\underline{Pr}\ (A), \overline{Pr}(A))$, in sense of Pawlak a rough set is an approximation of a set in terms of a pair of sets which gives the lower and upper approximation of original set .The set F_A give rise to an approximation space in Pawlaks' sense .

Note 4.5.1.6:

i. The initial description of the object of soft set has an approximate nature and we do not need the nation of exact solution

ii. The absence of any restriction on the approximate description in set theory makes it practically very suitable and easy to apply .

iii. According to the note 3.1.7 we deal with soft fuzzy sets as a soft sets and we apply it in this section .

The following definitions are related with our study.

Definition4.5.1.7:

Let $\tilde{R}$ be a fuzzy subset of $X \times Y$ and $\tilde{R}$ is defined as a fuzzy relationship from X to Y. We write X$\overset{\tilde{R}}{\rightarrow}$Y and $\tilde{R}(x,y)$ denotes the degree of corresponding between x and y based on the relationship $\tilde{R}$.

Let $F(X \times Y)$denotes the family of fuzzy relationship on X to Y . The set

$\tilde{R}_\alpha = \{ (x,y) \in X \times Y \mid \tilde{R}(x,y) \geq \alpha\} \subset X \times Y$ is defined as cut-set if $\tilde{R} \subset (X \times Y)$ for $\alpha \in [0,1]$. Then $\tilde{R}_\alpha$ denotes a relationship from X to Y and $\tilde{R}_\alpha$ is called $\alpha - cut$ relationship.

X,Y initial universal set, R defined from X to Y can be decided as a subset of $X \times Y \ \forall \ (x,y) \in X \times Y$

Definition4.5.1.8:

Let U be an initial universal set and E be a set of parameters Let P(U) denotes the power set of U . Let A⊂E, A pair (F,A) is called a soft fuzzy set over U. where F is a mapping given by F:A→P(U) and F(x)= [y∈ $U \mid \tilde{R}(x,y) \geq \alpha$] x∈ A , $y \in U$, $\alpha \in [0,1]$ (Here $\tilde{\tilde{R}}$ denotes fuzzy relationship between E and U.

Definition 4.5.1.9 :

Let $S = (U, R)$ be soft fuzzy approximation space with relation R and τ_S is the topology associated to S. Then the triple $M = (U, R, \tau_S)$ is called a topologized soft fuzzy approximation space.

Definition 4.5.10 :

Let $M_S = (U, R, {}_S\tau_f)$ be a topologized soft fuzzy approximation space and ${}_SA_f \subseteq U$. The lower soft fuzzy approximation (resp. upper soft approximation) of ${}_SA_f$ are:

$F_*(x) = R_*(F(x))$is lower soft fuzzy approximation and $F^*(x) = R^*(F(x))$ is upper soft fuzzy approximation.

Definition4.5.1.11: A soft fuzzy probability space over universal U is denoted by (U,A,pr) where A is a soft $\sigma - algabra$ over U and pr is the soft probability measure over the soft set A

Example 4.5.1.12 :

Let $U = \{a,b,c\} and\ R = \{(a,a), (a,b), (b,b), (b,c)\}$. $Then\ U/R = \{\{a,b\},\{b,c\}\}$.

Let $N = (U, R)$ be an approximation space and τ_N is the topology associated to N. Thus,

$\tau_N = \{U, \emptyset, \{b\}, \{a,b\}, \{b,c\}\}$,

If A={a,c} , $\tau_N^c = \{\emptyset, U, \{a,c\}, \{c\}, \{a\}\}$

then $R^*(A) = cl.A$

$= U \cap \{a,c\} = \{a,c\}$, hence $\overline{Pr}(A) = pr(cl.A) = \frac{2}{3}$

Now $R_*(A) = int.A = \emptyset$ therefor $\underline{pr}(A) = pr(int.A) = 0$

Example 4.5.1.13 :

Let $M = (X, R, \tau_M)$ be a topological approximation space such that

$X = \{1,2,3,4\}$.

Let R be a binary relation defined on U such that

$$R = \{(1,1), (2,2), (3,2), (3,4)\}.$$

$X/R = \{\{1\}, \{2\}, \{2,4\}\}$. Then the sub base $S = \{\{1\}, \{2\}, \{2,4\}\}$. Since X/R is a sub base of τ_M , and the base $B = \{\emptyset, X, \{1\}, \{2\}, \{2,4\}\}, thus$ $\tau_M = \{X, \emptyset, \{1\}, \{2\}, \{1,2\}, \{2,4\}, \{1,2,4\}\}$

Proposition 4.5.1.14 :

Let $M_S = (U, R, \tau_S)$ be a topologized soft fuzzy approximation space. If A and B are two soft fuzzy subsets of U, then

1) $R_*A \subseteq A \subseteq R^*A$.

2) $R_*\emptyset = R^*\emptyset = \emptyset$ and $R_*X = R^*X = X$.

3) $R^*(A \cup B) = R^*A \cup R^*B$.

4) $R_*(A \cap B) = R_*A \cap R_*B$.

5) If $A \subseteq B$, then $R^*A \subseteq R^*B$.

6) If $A \subseteq B$, then $R_*A \subseteq R_*B$.

7) $R_*(A \cup B) \supseteq R_*A \cup R_*B$.

8) $R^*(A \cap B) \subseteq R^*A \cap R^*B$.

9) $R_*(A)^c = (R^*A)^c$.

10) $R^*(A)^c = (R_*A)^c$.

11) $R_*\ R_*A = R_*A$.

12) $R^*\ R^*A = R^*A$.

Proof. By using soft fuzzy interior and soft fuzzy closure properties which are in chapter three the proof is clear.

Example 4.5.1.15 :

Let $K = (U, R, \tau_K)$ be a topological approximation space such that $U = \{a, b, c, d\}$and $U/_R = \{\{c\}, \{b, d\}\}$then the sub base $S = \{\{c\}, \{b, d\}\}$ the base

B is $B = \{U, \emptyset, \{c\}, \{b, d\}\}$and $\tau_K = \{U, \emptyset, \{c\}, \{b, d\}, \{b, c, d\}\}$

Lea $A = \{a, b, c\}$,now $R_*R_*A = int(int.A) = \{\{c\} \cup \emptyset\} = \{c\} = int.A$

And $R^*R_*A = cl(int.A) = cl(\{c\})$

$\tau_K^c = \{U, \emptyset, \{a, b, d\}, \{a, c\}, \{a\}\}$

Therefor $cl(\{c\}) = U \cap \{a, c\} = \{a, c\}$.

Exercises

1. Given an geometric and probabilistic interpretion for a fuzzy set .
2. Let M be a soft σ – **algebra** over X . Show that , if $M_e = \{ F(e); F_E \in M , e \in E \}$, then M_e is a σ – algebra over X .
3. A company wants to fill a job . There are seven candidates who form the set of alternatives , $X = \{ x_1, x_2 , ... , x_7 \}$. The hiving committee determine a set of parameters $E = \{ e_1 , e_2 , ..., e_5 \}$ where the parameters e_i , i=1,..,5 are experience , young age , good speaking , friendly and computer knowledge respectively . According to the constraints to a choosen subset $A = \{ e_2 , e_3 , e_4 \}$ list the steps to select a candidate for the job as an application on fuzzy soft sets .
4. Apply soft topology in GIS studied of predications as flood and rates of water tables .
5. Let $M = (U , R , T_N)$ be a topological space such that $U =\{ a, b, c, d \}$ and $U/R = \{\{d\} , \{ a , b \}\}$
   a- Determine the subbase , the base and T_M
   b- Let $A = \{ a , b , d \}$ find R^*R^*A and R_*R^*A .

Chapter Five

Applications in Uncertainty

Using Different Mathematical Tools

5. Preliminaries:

In this chapter, applications in different types of uncertainty are presented through illustrative examples in medical diagnoses, active teaching related with total quality and in reliability for many Systems of systems by using different mathematical tools as fuzzy sets , vague sets , intuitionistic fuzzy sets, type-2 fuzzy set and soft intuitionistic fuzzy sets more over several applications in different types fault tree analysis and confidence interval .

5.1 Types of Uncertainty with Applications

Three types of uncertainty are recognized in which the measurement of uncertainty is well established

1.non specificity (or imprecision) which is connected with size (cardinalities) of relevant sets of alternatives

2. fuzziness (or vagueness) which is result from imprecise boundaries of fuzzy sets)

3. Strife (or discord) which resemble conflicts through the different sets of alternative

Klir and wieman classify uncertainty into two major categories as below :

1. Fuzziness which deals with information that is indistinct.
2. Ambiguity which deals with multiplicity.

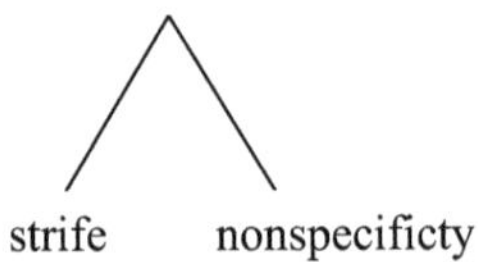

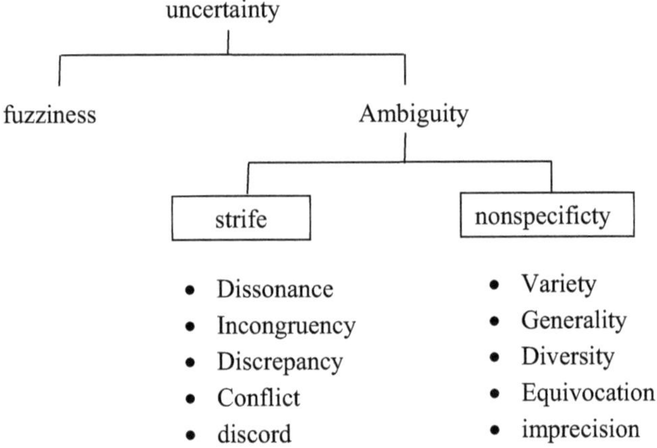

Soft Generalized Vague Sets An application in Medical Diagnosis 5.1.1:

We recall some definitions related with the study and the definition of soft set and soft fuzzy set are presented in section 1.6 and 3.1.1 respectively

Definition 5.1.1.1:

Let be the universal of discourse. A vague set $\bar{v}$ over U is characterized by a truth membership function $t_{\bar{v}}$ $\bar{v}$ a false membership function $f_{\bar{v}}$, $f_{\bar{v}}$:]. If generic element of U is denoted by then the lower bound on the membership grade of derived from evidence for is denoted by $t_{\bar{v}}(x_i)$and the lower bound on the negation of x_i is denoted by $f_{\bar{\mathrm{v}}}(x_i)$. $t_{\bar{\mathrm{v}}}(x_i)$ and $f_{\bar{\mathrm{v}}}(x_i)$ both associate a real number in [0,1] with each point x_i in X , where $t_{\bar{\mathrm{v}}}(x_i) + f_{\bar{\mathrm{v}}}(x_i) \leq 1$

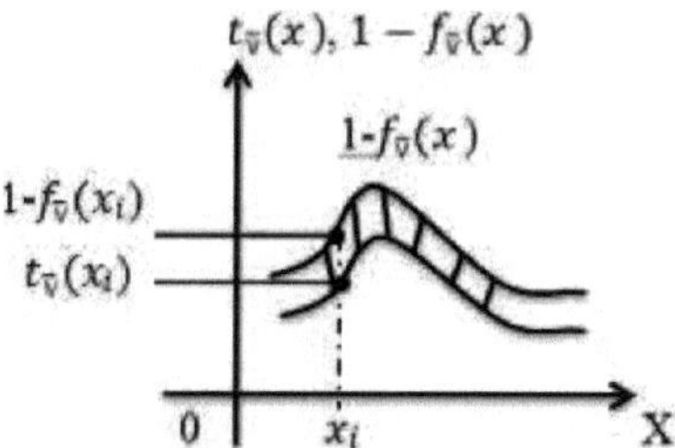

Figure (5-1) . A vague Set.

Definition 5.1.1.2 :

Let U the universal set and E be the set of parameters , let A ⊆E and F:A→V(U) and ∝ be a vague subset of A i.e α: A→[0,1], where V(U) the collection of all vague subsets of U. Let $\tilde{F}_\alpha$:A→ V(U) × [0,1]be a function defined as follows $\tilde{F}_\alpha(\mathrm{a}) = \{\tilde{F}_\alpha(\mathrm{a}) = \{x, t_{F(x)}, 1 - f_{F(x)}\}, \alpha(\mathrm{a})\}$ Then $\tilde{F}_\alpha$ is called the generalized vague soft set over (U,E)

Definition 5.1.1.3 :

Let $\tilde{F}_\alpha$ and $\tilde{G}_\beta$ be two generalized vague soft sets our (U,E). then the generalized vague soft relation R from $(\tilde{F}, \alpha)$ **to** $(\tilde{G}, \beta)$ is the function R: A × B→V(U) × [0,1] defined by R(a,b)$\tilde{\subseteq}$ $(\tilde{F}_\alpha(a) \tilde{\cap} \tilde{G}_\beta(b))$ $\forall (a,b) \in A \times B$.

Now we introduced the definition of soft generalized vague set is and we define some operations on it as below.

Definition 5.1.1.4:

Let $\tilde{R}_\alpha(\tilde{R}_{t_{\bar{v}}}, \tilde{R}_{f_{\bar{v}}})$be a vague subset of $X \times Y$ and $\tilde{R}_\alpha$ is defined as vague relationship from X to Y and $\tilde{R}(x,y)$ denotes the degree of correspondence between x and y based on the relationship $\tilde{R}$. Let V($X \times Y$) represents the family of vague relationship from X to Y. The set $\tilde{R}_\alpha = \{(x,y) \in X \times Y: \tilde{R}_{t_{\bar{v}}}(x,y) \geq \alpha \; and \; \tilde{R}_{f_{\bar{v}}}(x,y) \leq \alpha\} \subset X \times Y$ is defined as $\alpha - cut$ set if $\tilde{R} \in$ V($X \times Y$)for $\alpha \in [0,1]$.

Definition 5.1.1.5:

Let X be an initial set and E be a set of parameters. Let $P(X)$ represents the power set of X . Let A$\subset E$. A pair (V,A) is called a soft generalized vague

set over X . V is a mapping given by $V: A \rightarrow P(X)$ and

V(x)={y$\in X| \ (x, y) \in R_\alpha, x \in A, y \in X, \alpha \in [0,1]$}

Definition 5.1.1.6:

For two soft generalized vague sets $(\mathrm{G}, A_\alpha)_{\tilde{R}}$ and $(\mathrm{H}, B_\beta)_{\tilde{R}}$ over common universal X .

$(\mathrm{G}, A_\alpha)_{\tilde{R}}$ is a soft generalized vague subset of the soft generalized vague set $(\mathrm{H}, B_\beta)_{\tilde{R}}$, (briefly , (G,$A_\alpha$) $\tilde{\subseteq}$ $(\mathrm{H}, B_\beta)_{\tilde{R}}$) it

(i) $\alpha \subseteq \beta$

(ii) A$\subseteq$B

(iii) $\forall \alpha \in A, G(a)$ is a soft generalized vague set

Two soft generalized vague sets $(\mathrm{G}, A)_{\tilde{R}}$ and $(\mathrm{H}, B)_{\tilde{R}}$ over common universal X are equal if $(\mathrm{G}, A)_{\tilde{R}} \subset (\mathrm{H}, B)_{\tilde{R}}$
and $(\mathrm{G}, B)_{\tilde{R}} \subset (\mathrm{H}, A)_{\tilde{R}}$

Definition 5.1.1.7:

Let $(\mathrm{V}, \mathrm{A})_{\tilde{R}}$ a soft generalized vague set over the universal X is an empty soft generalized vague set (ϕ) if $\forall x \in A$, V(x)=ϕ (null or empty soft generalized vague set) .

Definition 5.1.1.8:

Let $(\mathrm{G}, A)_{\tilde{R}}$ and $(\mathrm{H}, B)_{\tilde{R}}$ are two soft generalized vague sets over the common universal X . Their union is the soft generalized set $(\mathrm{P}, C)_{\tilde{R}}$ where C=A$\cup B$ and e$\in C$

$$\mathrm{P(e)}=\begin{cases} G(e) & e \in A-B \\ H(e) & e \in B-A \\ G(e) \cup H(e) & e \in A \cap B \end{cases}$$

and we write the union as below:

$(\mathrm{G}, A)_{\tilde{R}} \ \tilde{\cup} \ (\mathrm{H}, B)_{\tilde{R}} = (\mathrm{P}, C)_{\tilde{R}}$

Definition 5.1.1.9:

The intersection of two soft generalized vague sets $(\mathrm{G}, A)_{\tilde{R}}$ and $(\mathrm{H}, B)_{\tilde{R}}$ over the common universal X is the soft generalized vague set $(\mathrm{Q}, C)_{\tilde{R}}$ where C=A∩B and $\forall e \in C$

$$Q(e)=\begin{cases} G(e) & e \in A-B \\ H(e) & e \in B-A \\ G(e) \cup H(e) & e \in A \cap B \end{cases}$$

and write the intersection as below: $(\mathrm{G}, A)_{\tilde{R}} \tilde{\cap} (\mathrm{H}, B)_{\tilde{R}} = (\mathrm{Q}, C)_{\tilde{R}}$

OR and AND operations:

We introduce the operatoins AND , OR on soft generalized vague set respectively .

Definition 5.1.1.10:

If $(\mathrm{G}, A)_{\tilde{R}}$ and $(\mathrm{H}, B)_{\tilde{R}}$ be two soft generalized vague sets then "$(\mathrm{G}, A)_{\tilde{R}}$ and $(\mathrm{H}, B)_{\tilde{R}}$" is a soft generalized vague set.

Denoted by: $(\mathrm{G}, A)_{\tilde{R}} \wedge (\mathrm{H}, B)_{\tilde{R}}$

And defined by :

$(\mathrm{G}, A)_{\tilde{R}} \wedge (\mathrm{H}, B)_{\tilde{R}} = (\mathrm{P}, \mathrm{A} \times \mathrm{B})_{\tilde{R}}$ where

$\mathrm{P}(\alpha, \beta) = \mathrm{G}(\alpha) \cap H(\beta) \quad \forall(\alpha, \beta) \in A \times B$

Definition 5.1.1.11:

If $(\mathrm{G}, A)_{\tilde{R}}$ and $(\mathrm{H}, B)_{\tilde{R}}$ are two soft generalized vague sets then $(\mathrm{G}, A)_{\tilde{R}}$ OR $(\mathrm{H}, B)_{\tilde{R}}$ is a soft generalized vague set denoted as follows: $(\mathrm{G}, A)_{\tilde{R}} \vee (\mathrm{H}, B)_{\tilde{R}}$ and defined by:

$(\mathrm{G}, A)_{\tilde{R}} \vee (\mathrm{H}, B)_{\tilde{R}} = \llbracket (Q, AXB) \rrbracket \_R$̃

where

$\mathrm{Q}(\alpha, \beta) = \mathrm{G}(\alpha) \cup H(\beta) \quad \forall(\alpha, \beta) \in A \times B$

Example 5.1.1.12:

Suppose that is a patient having certain visible symptoms is infected by Ameobiasis disease , which is characterized by the infection of human large? intestine by the parasite Entamoeba histolytica . This disease so called Amebic dysentery . An investigation in a clinical labrotorary is performed to diagnosis whether the patient is infected by this disease or not. The symptoms of the disease is charactized by : mucoid stool , diarrhea , vomiting , abdominal pain , abdominal cramp, and may be reach to (bloody diarrhea). Let the universal set X consists of two elements (cases) , infected denoted by c_1 and non infected denoted by c_2 i.e. $X=\{c_1, c_2\}$

i.e. $\mathrm{X}=\{c_1, c_2\}$, the set of parameters E is the set of certain approximations determined by the clinical laboratory symptoms diagnosis

Let E = {e_1 (macoid stool) , e_2 (diarrhea) , e_3(vomiting) , e_4(abdominal pain) , e_5(abdominal cramp) , e_6(bloody diarrhea)}

To determine whether the patient suffering from Ameobiasis disease or not . we follow the following steps:

1. Suppose the relation $\tilde{R}_\alpha$ is :

$$\tilde{R}_\alpha = (0.5\,,0.8)/(c_1\,,e_1) + (0.6\,,0.7)/(c_1,e_2) + (0.4\,,0.6)/(c_1\,,e_3) + (0.4\,,0.7)/(c_1,e_4) + (0.5\,,0.5)/(c_1\,,e_5) + (0.7\,,0.9)\,/(c_1,e_6) + (0.3\,,0.4)/(c_2\,,e_1) + (0.1\,,0.9)/(c_2,e_2) + (0.5\,,0.6)/(c_2\,,e_3) + (0.0\,,0.8)/(c_2,e_4) + (0.1\,,0.9)/(c_2\,,e_5) + (0.2\,,0.8)/(c_2,e_6)$$

Let $\alpha(e_1) = 0.3$, $\alpha(e_2) = 0.5$, $(e_3) = 0.5$, $\alpha(e_4)$ =0.6 , $\alpha(e_5) = 0.2, \alpha(e_6) = 0.7$

The soft generalized vague set as a collection of vague approximation as below :

$F_\alpha(e_1) = \{(c_1\,,0.5\,,0.8), (c_2\,,0.3\,,0.4), 0.3\}$
$F_\alpha(e_2) = \{(c_1\,,0.6\,,0.7), (c_2\,,0.1\,,0.9)\,,0.5\}$
$F_\alpha(e_3) = \{(c_1\,,0.4\,,0.6), (c_2\,,0.5\,,0.6)\,,0.4\}$
$F_\alpha(e_4) = \{(c_1\,,0.4\,,0.7), (c_2\,,0\,,0.8)\,,0.6\}$
$F_\alpha(e_5) = \{(c_1\,,0.5\,,0.5), (c_2\,,0.1\,,0.9)\,,0.2\}$
$F_\alpha(e_6) = \{(c_1\,,0.7\,,0.9), (c_2\,,0.2\,,0.8)\,,0.7\}$

2. The truth- membership function $\check{t}_{c_k} = a_i + b_i - a_i b_i$

   Where $a_i = t_{c_k}(e_i)$, $i = 1,2,\dots,6$ K=1,2

   and $b_i = \alpha(e_i)$

   and The false – membership function

   $1 - \check{f}_{c_k}(e_i) = d_i f_i$ where $d_i = 1 - f_i$

   and $f_i = \alpha(e_i)$

To determine the high truth-membership value and low false – membership value we use $\check{t}_{c_k}(e_i)$ and 1- $\check{t}_{c_k}(e_i)$ respectively to get the soft vague set $F_\alpha(e_i)$ below:

$F_\alpha(e_1) = \{(c_1\,,0.65\,,0.24), (c_2\,,0.51\,,0.12), 0.3\}$
$F_\alpha(e_2) = \{(c_1\,,0.80\,,0.35), (c_2\,,0.55\,,0.45), 0.5\}$
$F_\alpha(e_3) = \{(c_1\,,0.56\,,0.24), (c_2\,,0.66\,,0.24), 0.4\}$
$F_\alpha(e_4) = \{(c_1\,,0.76\,,0.42), (c_2\,,0.6\,,0.8), 0.6\}$
$F_\alpha(e_5) = \{(c_1\,,0.6\,,0.1), (c_2\,,0.28\,,0.18), 0.2\}$

$F_\alpha(e_6) = \{(c_1, 0.91, 0.63), (c_2, 0.76, 0.56), 0.7\}$

3. The representation of the truth-membership function and the false – membership function are as in the tables(5- 1) and (5-2) below respectively

Table (5-1) :Representation of the truth-membership function

	e_1	e_2	e_3	e_4	e_5	e_6
c_1	0.65	0.80	0.56	0.76	0.6	0.91
c_2	0.51	0.55	0.66	0.6	0.28	0.76

Table (5-2) :Representation of the false-membership function

	e_1	e_2	e_3	e_4	e_5	e_6
c_1	0.24	0.35	0.24	0.42	0.1	0.63
c_2	0.12	0.45	0.24	0.48	0.18	0.56

4. Calculating the comparison table which is a square table (i.e the number of rows and columns are equal) and both are labeled by the object name of the universal X= $\{c_1, c_2\}$. The entries k_{ij} represent the number for which c_i greater than or equal to the value of $c_j$as shown in the table (5-3) , (5-4) below :

Table (5-3) : Comparison table of the truth-membership function

	c_1	c_2
c_1	6	5
c_2	1	6

Table (5-4) : Comparison table of the false-membership function

	c_1	c_2
c_1	6	3
c_2	4	6

5. Calculate the truth-membership score and false-membership score As shown in tables (5-5) and (5-6) below:

Table (5-5) : truth-membership score table

Row sum (m)	Column sum (n)	Truth Membership score(m-n)
11 7	7 11	4 -4

Table (5-6) : false-membership score table

Row sum (m)	Column sum (n)	False membership score(m-n)
9 10	10 9	-1 1

Find the difference score = truth-membership score - false-membership score

Table (5-7) : false-membership score table

	truth-membership score	false-membership score	Score difference
c_1 c_2	4 -4	-1 1	5 -5

6. Finding the maximum score :

From the table (5-7) above , the maximum score is 5 which means that the patient infected by ameobiasis disease .

An Application in Total Quality for development of Active Teaching technique by Using Fuzzy Soft Sets5.1.2

In this section, we need the definitions and properties of fuzzy sets, soft sets and fuzzy soft sets which are mentioned in sections 1.2, 1.4 and 2.2

Example as an application 5.1.2.1

The following example is an application in total quality for active teaching

As an application in total quality for active teaching suppose F is the fuzzy soft set represent the teaching rendering of seven instructors in the universal U where

U=$\{i_1, i_2, i_3, i_4, i_5, i_6, i_7\}$ and E be the set of parameters representing specific skills for the instructors.

E=$\{e_1$ $(using\ smart\ board)$,
$e_2(Rein\ forcement), e_3(explanation), e_4(using\ illustrative\ examples)$,

$e_5(take\ care\ with\ the\ individual\ differences), e_6(probing\ questions)$,

e_7(introducing a lesson)$\}$

Let $A \subseteq E$ where A=$\{e_1, e_2, e_4, e_7\}$ be the standard parameters to choice on instructor.

Let $\mu : A \to [0,1]$ be a fuzzy subset of A defined as follows :

$\mu(e_1) = 0.3$, $\mu(e_2) = 0.4$, $(e_4) = 0.2$, $\mu(e_7) = 0.7$. and defined a mapping

$F_\mu : E \to P(U) \times [0,1]$

1. consider the fuzzy soft set F_μ as the following :

$F_\mu(e_1) = (\{i_1/\ 0.3, i_2/0.2, i_3/0.4, i_4/0.4, i_5/0.5, i_6/1, i_7/0.5\}0.3)$

$F_\mu(e_2) = (\{i_1/\ 0.5, i_2/1\ , i_3/0.4, i_4/0.2, i_5/0.7, i_6/0.3, i_7/0.7\}0.4)$

$F_\mu(e_4) = (\{i_1/\ 0.2, i_2/0.6, i_3/0.3, i_4/0.5, i_5/0.8, i_6/0.2, i_7/0.3\}0.2)$

$F_\mu(e_7) = (\{i_1/0.5, i_2/0.7, i_3/0.7, i_4/0.8, i_5/0.4, i_6/0.8, i_7/0.6\}0.7)$

We put F_μ in the table below:

Table (5-8) : Table of F_μ

	e_1	e_2	e_4	e_7
i_1	0.3	0.5	0.2	0.5
i_2	0.2	1	0.6	0.7
i_3	0.4	0.4	0.3	0.7
i_4	0.4	0.2	0.5	0.8
i_5	0.2	0.7	0.8	0.4
i_6	1	0.3	0.2	0.8
i_7	0.5	0.7	0.3	0.6

2- Finding the complement of F_μ from the relation $F_\mu{}^c = 1 - F_\mu$ as shown in the table(5- 9):

Table(5- 9) :Table of $F_\mu{}^c$

	e_1	e_2	e_4	e_7
i_1	0.7	0.5	0.8	0.5
i_2	0.8	0	0.4	0.3
i_3	0.6	0.6	0.7	0.3
i_4	0.6	0.8	0.5	0.2
i_5	0.8	0.3	0.2	0.6
i_6	0	0.7	0.8	0.2
i_7	0.5	0.3	0.7	0.4

2. Multiplying every entry of the table of F_μ by the corresponding values of $\mu(e)$ with the row sum which represent the membership score and denoted it by a.

Table(5- 10): Membership Score Table

	$e_1(0.3)$	$e_2(0.4)$	$e_4(0.2)$	$e_7(0.7)$	Row sum a
i_1	0.09	0.20	0.04	0.35	0.68
i_2	0.06	0.4	0.12	0.49	0.62
i_3	0.12	0.16	0.06	0.49	0.83
i_4	0.12	0.08	0.10	0.56	0.86
i_5	0.06	0.28	0.16	0.28	0.78
i_6	0.30	0.12	0.04	0.56	1.02
i_7	0.15	0.28	0.06	0.42	0.91

4- We repeat the step (3) for the table of F_{μ}^{c} with the row sum which represent the non-membership score and denoted by b:

Table (5-11): Non – Membership Score Table

	$e_1(0.7)$	$e_2(0.6)$	$e_4(0.8)$	$e_7(0.3)$	Row sum b
i_1	0.49	0.30	0.64	0.15	1.58
i_2	0.56	0	0.32	0.09	0.97
i_3	0.42	0.36	0.56	0.09	1.43
i_4	0.42	0.48	0.40	0.06	1.36
i_5	0.56	0.18	0.16	0.18	1.8
i_6	0	0.42	0.64	0.06	1.12
i_7	0.35	0.18	0.56	0.12	1.21

5- Finally we calculate a+b-ab and takes the minimum score which determine the instructor has the best teaching rendering among all the other instructors under consideration.

Table (5-12): Table of a+b-ab

	a	b	a+b–ab
i_1	0.68	1.58	2.26 – 1.07=0.19
i_2	0.62	0.97	1.59 – 0.60 = 0.99
i_3	0.63	1.43	1.06 – 0.90 =0.16
i_4	0.86	1.36	1.22 – 1.16 = 0.06
i_5	0.78	1.80	1.58 – 1.40 = 0.18
i_6	1.02	1.12	2.14 – 1.14 =1
i_7	0.91	1.22	1.13 – 1.11 = 0.02

Hence from table (5-12) i_7 valid the best teaching rendering since min (a+b-ab)=0.02 .

5.2 Uncertain Reliability Applications by using different mathematical tools

Uncertain reliability analysis is a tool to deal with system reliability via uncertainty theory.

Structure function 5.2.1:

In crisp reliability, any element has two states : working and failure . we denote the state of elements i by the Boolean variables

$$x_i = \begin{cases} 1 & if\ element\ i\ works \\ 0 & if\ element\ i\ fails \end{cases} \quad ......(5\text{-}1)$$

$i = 1,2, ..., n$, we also denote the state of the system by the Boolean variable

$$X = \begin{cases} 1 & if\ system\ i\ works \\ 0 & if\ system\ i\ fails \end{cases} \quad ....(5\text{-}2)$$

Note 5.2.2 .

The state of the system is completely determined by the state of its elements via the so called structure function.

Definition 5.2.3:.

Assume that X is a Boolean system containing elements $x_1, x_2, ..., x_n$. A Boolean function f is called a structure function of X if

$$\mathrm{X} = 1 \Leftrightarrow f(x_1, x_2, ..., x_n) = 1 ... . (5-3)$$

$$X = 0 \Leftrightarrow f(x_1, x_2, \dots, x_n) = 0 \dots\dots\dots (5-4)$$

whenever f is indeed the structure function of the system.

i. For a series system, the structure function is a mapping from $\{0,1\}^n$ to $\{0,1\}$ *i.e* (AND)

$$f(x_1, x_2, \dots, x_n) = x_1 \wedge x_2 \wedge \dots \wedge x_n \dots\dots\dots (5\text{-}5)$$

Figure (5-2). A series system

The series System is called line topology since if any component fails then all the line fails. See [124,126]

ii. For a parallel system, the structure function is a mapping from $\{0,1\}^n$ to $\{0,1\}$ *i.e* (OR)

$$f(x_1, x_2, \dots, x_n) = x_1 \vee x_2 \vee \dots \vee x_n \dots\dots\dots (5\text{-}6)$$

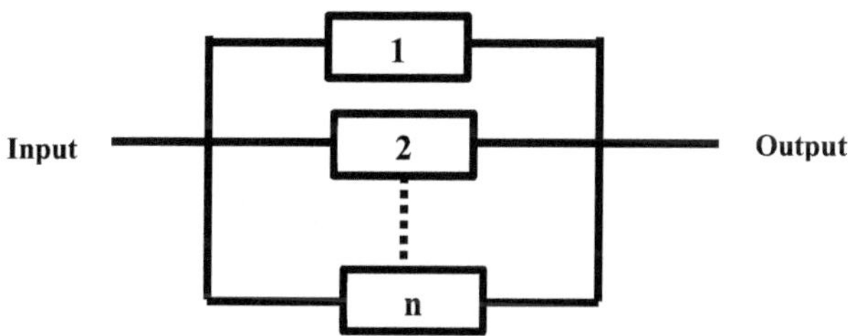

Figure(5- 3). A parallel system

iii. K-out–of-n system that works whenever at least K of the n elements work , the structure function is a mapping from $\{0,1\}^n$ to $\{0,1\}$ structure function $f(x_1, x_2, \dots, x_n) = K - \max\{x_1, x_2, \dots, x_n\}$....(5-7)

Note that when K=1 it is parallel system when K=n it is series system

$$f: \{0,1\}^n \to \{0,1\}$$

Reliability Index 5.2.4:

It is an uncertain measure that the system works the element in a Boolean system is usually represented by a Boolean uncertain variable. i.e,

$$\varepsilon = \begin{cases} 1 & \text{with uncertain measure } a \\ 0 & \text{with uncertain measure } 1 - a \end{cases} \dots\dots\dots (5\text{-}8)$$

In this case we will say ε is an uncertain element with reliability, a reliability index(in short RI) is defined as the uncertain measure denoted by $\mathcal{M}$ that the system is working .

Definition 5.2.5:

Assume a Boolean system has uncertain elements $\varepsilon_1, \varepsilon_2,..., \varepsilon_n$ and a structure function f . Then the reliability index *(RI)*is the uncertain measure that the system is working, *i.e,*

$RI = \mathcal{M}\{f(\varepsilon_1, \varepsilon_2,..., \varepsilon_n) = 1\}$............*(5-9)*

System Systems 5.2.6:

(a). Series System : consider a serious system having independent uncertain elements $\varepsilon_1, \varepsilon_2,..., \varepsilon_n$ with reliabilities $: a_1, a_2,..., a_n$ respectively .

From (i) the structure function $f(x_1, x_2, ..., x_n) = x_1 \wedge x_2 \wedge, ..., \wedge x_n$

$= \min(x_1, x_2, ..., x_n)$............(5-10)

it follows from the reliability index that the reliability index is

$RI = \mathcal{M}\{f(\varepsilon_1 \wedge \varepsilon_2, \wedge ..., \wedge \varepsilon_n)\} = 1$

$= a_1 \wedge a_2 \wedge, ..., \wedge a_n$ ....(5-11)

(b). Parallel System : Consider a parallel system having independent uncertain elements $\varepsilon_1, \varepsilon_2,..., \varepsilon_n$ with reliabilities $: a_1, a_2,..., a_n$ respectively .

From (ii) the structure function $f(x_1, x_2, ..., x_n) = x_1 \vee x_2 \vee ..., \vee x_n$

it follows from the reliability index that the reliability index is

Reliability Index RI= $\mathcal{M}\{f(\varepsilon_1 \vee \varepsilon_2, \vee ..., \vee \varepsilon_n)\} = 1$

$= a_1 \vee a_2 \vee ... , \vee a_n$ ....(5-12)

We can write the structure function for parallel system as complement of series system i.e. $f(x_1, x_2, ..., x_n) = 1 - [(1 - x_1)(1 - x_2) ..., (1 - x_n)] = \max(x_1, x_2, ..., x_n)$ ....(5-13)

(c). k-out-of-n system : Consider a k-out-of-n system having independent uncertain elements $\varepsilon_1, \varepsilon_2,..., \varepsilon_n$ with reliabilities $: a_1, a_2,..., a_n$ respectively .

From (iii)the structure function has a Boolean form

$f(x_1, x_2, ..., x_n) = k - \max\{x_1, x_2, ..., x_n\}$..........(5-14)

it follows from the reliability index that the reliability index is

k th largest value of $a_1, a_2,..., a_n$

i.e RI = k- max$\{a_1, a_2,..., a_n\}$ ....(5-15)

(d). General system :

Consider the following example taken from [95]

Example5.2.7:

Let a bridge system as shown in figure below

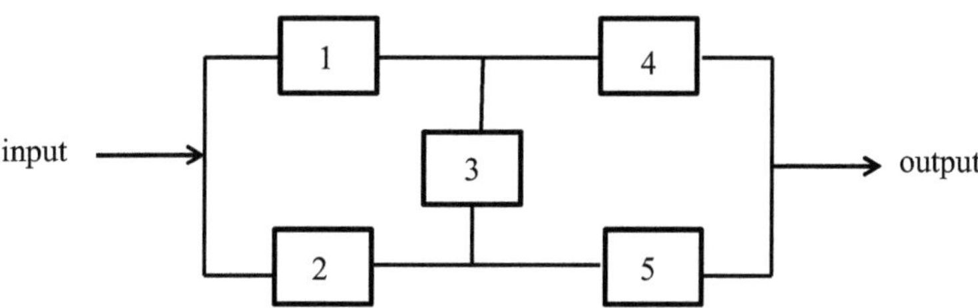

Figure (5-4) A Bridge System

It is consist of 5 independent uncertain elements whose states defined by $\varepsilon_1, \varepsilon_2, \varepsilon_3, \varepsilon_4$ *and* ε_5 . Assume each path works iff all elements on which is working and the system works iff there is a path of working element . Then the structure function of the bridge in figure above :

$$f(x_1, x_2, x_3, x_4, x_5) = (x_1 \wedge x_4) \vee (x_2 \wedge x_5) \vee (x_1 \wedge x_3 \wedge x_5) \vee (x_2 \wedge x_3 \wedge x_4) \text{.......} \quad (5\text{-}16)$$

By the Boolean system calculator : A function in the matlab uncertain tool assume the 5 independent uncertain elements have reliabilities 0.91 ,0.92 ,0.93 ,0.94 ,0.95 in uncertain measure . A run in the system calculator shows that the reliability index is = $\mathcal{M}\{f(\varepsilon_1, \varepsilon_2, ..., \varepsilon_5) = 1\} =$ 0.92 in an uncertain measure.

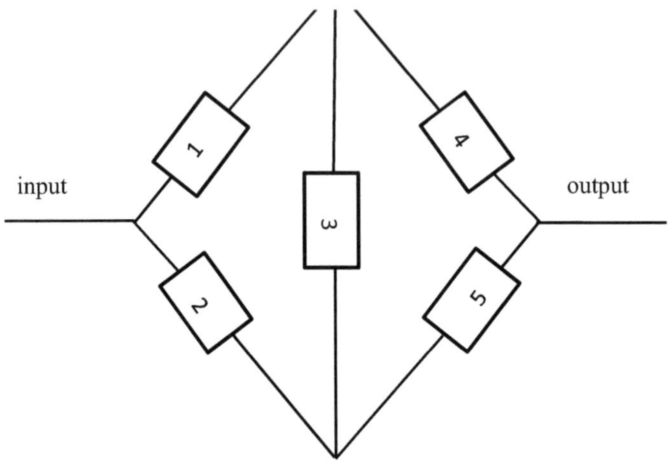

Figure (5-5) a . Bridge network of figure (5-4)

There exists four paths $P_1 = \{x_1, x_4\}, P_2 = \{x_2, x_5\}, P_3 = \{x_1, x_3, x_5\},$

$$and\ P_4 = \{x_2, x_3, x_4\}$$

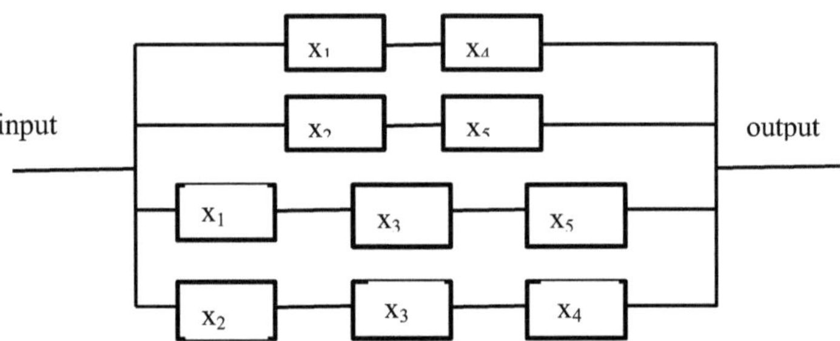

Figure (5-5) b. A network for figure (5-4)as parallel series system

The cut sets of the network of figure (5-4) are four cut sets

$$C_1 = \{x_1, x_2\},, C_2 = \{x_4, x_5\}, C_3 = \{x_1, x_3, x_5\},$$

$$and\ C_4 = \{x_2, x_3, x_4\}$$

It can easy to reduce the figure (5-5)a into series system or (line topology)via reduction method .

5.3 Using Triangular Fuzzy Numbers to Calculate the Fuzzy Reliability of the Network Bridge System

The steps are as follows:

1. Calculating the reliabilities for the four paths P_1, P_2, P_3 and P_4 where $P_1 = \{x_1, x_4\}, P_2 = \{x_2, x_5\}, P_3 = \{x_1, x_3, x_5\}$ and $P_4 = \{x_2, x_3, x_4\}$. The system reliability $\tilde{R}_s$ in terms of probability is $\tilde{R}_s = 1 - \{\tilde{P}_1, \tilde{P}_2, \tilde{P}_3, \tilde{P}_4\}$.
2. Assume that the triangular membership function for components C_1, C_2, C_3, C_4 and C_5 respectively of the network is the formula:

$$C_i(x) = \begin{cases} 0, x \leq a, c \leq x \\ \frac{x-a}{b-a}, a \leq x \leq b \\ \frac{c-x}{c-b}, b \leq x \leq c \end{cases} = (a, b, c) \ , \ i = 1,2,3,4,5 \quad \text{..........(5-17)}$$

For numerical calculating, we consider the following triangular numbers:

$$C_1(x) = \{0.25, 0.90, 0.95\} = \begin{cases} 0, x \leq 0.25, x \geq 0.95 \\ \frac{x - 0.25}{0.65}, 0.25 \leq x \leq 0.90 \\ \frac{0.95 - x}{0.5}, 0.90 \leq x \leq 0.95 \end{cases}$$

$$C_2(x) = \{0.30, 0.89, 0.97\} = \begin{cases} 0, x \leq 0.30, x \geq 0.97 \\ \frac{x - 0.30}{0.59}, 0.30 \leq x \leq 0.89 \\ \frac{0.97 - x}{0.08}, 0.89 \leq x \leq 0.97 \end{cases}$$

$$C_3(x) = \{0.42, 0.72, 0.92\} = \begin{cases} 0, x \leq 0.42, x \geq 0.92 \\ \frac{x - 0.42}{0.30}, 0.42 \leq x \leq 0.72 \\ \frac{0.92 - x}{0.20}, 0.72 \leq x \leq 0.92 \end{cases}$$

$$C_4(x) = \{0.20, 0.49, 0.78\} = \begin{cases} 0, x \leq 0.20, x \geq 0.78 \\ \frac{x - 0.20}{0.29}, 0.20 \leq x \leq 0.49 \\ \frac{0.78 - x}{0.29}, 0.49 \leq x \leq 0.78 \end{cases}$$

$$C_5(x) = \{0.45,0.82,0.89\} = \begin{cases} 0, x \leq 0.45, x \geq 0.89 \\ \frac{x - 0.20}{0.37}, 0.45 \leq x \leq 0.82 \\ \frac{0.78 - x}{0.07}, 0.82 \leq x \leq 0.89 \end{cases}$$

3. We calculate the reliabilities of the paths P_1, P_2, P_3 and P_4 let the reliabilities of the components are R_1, R_2, R_3, R_4 and R_5 respectively by using approximation of multiplication via finding the α-cuts, since the elements of each interval are positive numbers. So, the operation is simple.

Noted that the approximation of multiplication is used because the product is not a triangular fuzzy number and fuzzy complement to calculate $\widetilde{R_s}$

$$P_1 = R_1 R_4 = \{0.0500 \quad 0.4410 \quad 0.7410\},$$

$$R_{\overline{14}} = \{1 - 0.050, 1 - 0.441, 1 - 0.741\} = \{0.95, 0.559, 0.259\}$$

$$P_2 = R_{\overline{14}} R_2 R_5 = \{0.1283, 0.4080, 0.2236\},$$

$$R_{\overline{25}} = \{1 - 0.1350, 1 - 0.7298, 1 - 0.8633\} = \{0.8650, 0.2702, 0.1367\}$$

$$P_3 = R_{\overline{25}} R_{\overline{14}} R_1 R_3 R_5 = \{0.0389, 0.0803, 0.0275\},$$

$$R_{\overline{135}} = \{1 - 0.0473, 1 - 0.5314, 1 - 0.7779\} = \{0.9527, 0.4686, 0.2221\}$$

$$P_4 = R_{\overline{14}} R_{\overline{135}} R_{\overline{25}} R_2 R_3 R_4 = \{0.0197, 0.0222, 0.0055\}.$$

Now the product of paths complements $\widetilde{P_1}, \widetilde{P_2}, \widetilde{P_3} and\ \widetilde{P_4}$ is

$$\widetilde{P_1}.\widetilde{P_2}.\widetilde{P_3}.\widetilde{P_4} = (0.06, 0.82, 0.91)$$

We have $\widetilde{R_s} = 1 - (\widetilde{P_1}.\widetilde{P_2}.\widetilde{P_3}.\widetilde{P_4})$

$$\therefore\ \widetilde{R_s} = 1 - (0.06 * 0.82 * 0.91)$$

$$\widetilde{R_s} = 1 - (0.039862) = 0.970138$$

5.4. Fuzzy Probist and Profust System Reliability

In this section fuzzy probist and profust system reliability are discussed and the fuzzy probist system reliability for series systems and parallels systems are reformulated by using $\alpha - cut$ methods.

Fuzzy Probist and Profust 5.4.1:

Fuzzy set theory has been applied to reliability theory engineering with great success in the past two decades. The incorporating of the fuzzy set theoretic concepts into the multidisciplinary area of reliability theory has been done by modifying the basis assumptions underlying the definition of reliability of a component or system. Conventional reliability theory is based on, among others, the following two fundamental assumptions.

1) Binary state assumption: the system can only be in either of the two crisp states viz. fully functioning or fully failed.
2) Probability assumption: the system failure behavior is fully characterized by the probability measures.

Although, these two assumptions are often valid, they are not reasonable in a large variety of cases. This called for the incorporation of the concepts of fuzzy set theory into these assumptions . thus 1) and 2) are modified as follows:

1') Fuzzy state assumption: the system success and failure are characterized by fuzzy states. At any given time the system can be viewed as being in one of the two states to some extent. Thus, system failure is not defined in a binary way, but in a fuzzy way.

2') Possibility assumption: the system failure behavior is fully characterized by the possibility measures.

For the sake of simplicity, the conventional reliability theory is called "PROBIST reliability theory", since it is based on assumptions 1) and 2). When 1) is replaced by 1') and / or 2) is replaced by 2'), the resultant is called fuzzy reliability theory. Thus, fuzzy reliability theory manifests itself in three different forms viz. PROFUST reliability theory, POSBIST reliability theory, and POSFUST reliability theory.

Fuzzy reliability theory in its various forms found applications, especially in fault tree analysis, reliability optimization and risk analysis. However, fuzzy mathematical techniques can be successfully applied to conventional reliability theory, without taking recourse to any form of the fuzzy reliability theories. This was first demonstrated by Park who applied fuuzy optimization techniques to the problem of reliability apportionment for a simple two-component series system. Later, Dhingra and Rao and Dhingra worked on reliability and redundancy apportionment for a four-stage and a five-stage overspeed protection systems, respectively, using crisp and fuzzy multi-objective optimization approaches. To summarize, all the three works use the noncompensatory min operator as the aggregator. Park used a linear membership function for the goals, whereas Dhingra and Rao and Dhingra used linear membership functions for the constraints and nonlinear membership function for the reliability. All the three works do not make use of any of the random search based global optimization methods. Thus, the whole family of meta-heuristics that can efficiently solve highly nonlinear nonconvex mixed- integer optimization problems, is overlooked.

A probist system (component) is a system in the context of probist(conventional) reliability theory is based on:

1) Probability assumption: The system failure behavior is fully characterized in the context of probability measures.
2) Binary state assumption: The system at only one of the two crisp states (fully working or fully failed) .

Definition 5.4.2:

Probist reliability means reliability of probist system (component). It is probability in conventional sense. For various reasons , this probability can be fuzzy , in form of linguistic values or intervals. In either form. it can treated as a fuzzy number that in the probist system reliability R_s can be treated as a fuzzy number with membership function $\mu_{R_s}(x)$.

Fuzzy Probist System Reliability 5.4.3:

Suppose a probist system contains n components with reliabilities $R_1, R_2, \dots, R_n$ respectively and suppose that the system reliability R_s is precisely determined by

$$R_s = F(R_1, R_2, \dots, R_n) \text{ ....(5-18)}$$

where F called system reliability functin , $where R_1, R_2, \dots, R_n \in [0,1]$.

i. Consider a probist series system consists of n components, that R_s is a s follows:

$$R_s = R_1.R_2.\dots.R_n = \prod_{i=1}^{n} R_i \text{ ....(5-19)}$$

We suppose that $R_1, R_2, \dots, R_n$ are triangular fuzzy numbers denoted $R_i = (a_{i1}, a_{i2}, a_{i3}), i = 1,2, \dots, n,$ for both series and parallel system .Then by using the $\alpha - level\ set$ or called $\alpha - cut$ of triangular fuzzy number $\tilde{A} = (a_1, a_2, a_3)$ which is denoted by $\tilde{A}_\alpha$ see [80]

Now
$(R_i)_\alpha = [(a_{i2} - a_{i1})\alpha + a_{i1}, (a_{i3} - a_{i2})\alpha + a_{i3}], \forall\alpha \in [0,1]$..........(5-20)

We have:

$(R_\alpha)_S = \prod_{i=1}^{n}(R_i)_\alpha = \left[\prod_{i=1}^{n}\left((a_{i2} - a_{i1})\alpha + a_{i1}\right), \prod_{i=1}^{n}(-(a_{i3} - a_{i2})\alpha + a_{i3})\right]$

Alternatively, if we denote $R_i = (m_i - \alpha_i, m_i, m_i + \beta_i)$, then

$$R_s = \left[\prod_{i=1}^{n}(m_i - \alpha_i), \prod_{i=1}^{n} m_i, \prod_{i=1}^{n}(m_i + \beta_i)\right] \ldots\ldots\ldots\ldots (5-21)$$

ii. Consider a probist parallel system contains n components, R_s *is*

$R_s = 1 - \prod_{i=1}^{n}(1 - R_i)$ ....(5-22)

Since $R_1, R_2, \ldots, R_n$ are triangular fuzzy numbers, then so are $1 - R_1, 1 - R_2, \ldots, 1 - R_n$. $(R_i)_\alpha = [1,1] - \prod_{i=1}^{n}\{[1,1] - (R_i)_\alpha\}$

We have:

$(R_\alpha)_S = \{1 - \prod_{i=1}^{n}(R_i)_\alpha = [\prod_{i=1}^{n}(-(a_{i2} - a_{i1})\alpha + 1 - a_{i1}), 1 - \prod_{i=1}^{n}(-(a_{i3} - a_{i2})\alpha + 1 - a_{i3})]\}, \forall\alpha \in [0,1]$

Alternatively, if we denote $R_i = (m_i - \alpha_i, m_i, m_i + \beta_i)$, then

$$R_s = \left[1 - \prod_{i=1}^{n}(1 - (m_i - \alpha_i)), 1 - \prod_{i=1}^{n}(1 - m_i), 1 - \prod_{i=1}^{n}(1 - (m_i + \beta_i))\right] \ldots\ldots\ldots\ldots (5-23)$$

5.5 Analyzing Fuzzy System Reliability Based on Triangular Vague Set

In this section, a new method for analyzing fuzzy system reliability based on vague set theory is presented, where the reliabilities of the components of a system are represented by vague sets defined in the universal of discourse [0,1],by using the triangular vague number .

Definition 5.5.1: A vague number is a vague subset in the universal of discourse U that is both convex and normal.

Let us consider the triangular vague set $\tilde{A}$ shown in figure (5-6) where the triangular vague set $\tilde{A}$ can be parameterized by tuple

$$< [(a, b, c); \mu_1], [(a, b, c); \mu_2] >$$

For convenience ,the tuple $< [(a, b, c); \mu_1], [(a, b, c); \mu_2] >$ can be also abbreviated into $< [(a, b, c); \mu_1; \mu_2] >$ and $0 \le \mu_1 \le \mu_2 \le 1$.

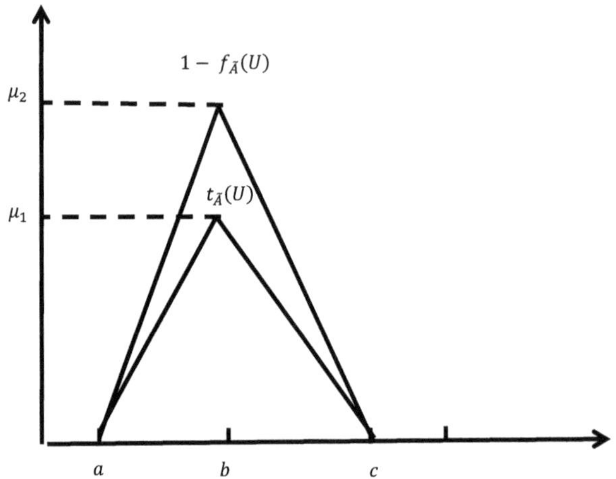

Figure (5-6). A Triangular vague set

Note 5.5.2: The arithmetic operations which are denoted by ,⊕,⊖,⊗,⊘ ,between the triangular fuzzy sets are used , for detail see [70]

Fuzzy Reliability for Series and Parallel Systems as in below 5.5.3:

i. **Series system:**

Consider a series system shown in Figure below, where the reliability $\tilde{R}_i$ of component i is represented by a vague set $< [(a_i, b_i, c_i); \mu_{i1}; \mu_{i2}] >$, where $0 \le \mu_{i1} \le \mu_{i2} \le 1$and $1 \le i \le n$. Then, the reliability $\tilde{R}$ of the series system can be evaluated as follows:

$\tilde{R} = \tilde{R}_1 \otimes \tilde{R}_2 \otimes ... \otimes \tilde{R}_n$ ....(5-24)

$=< [(a_1, b_1, c_1); \mu_{11}; \mu_{12}] > \otimes < [(a_2, b_2, c_2); \mu_{21}; \mu_{22}] > \otimes ...$

$\otimes < [(a_n, b_n, c_n); \mu_{n1}; \mu_{n2}] >$

$=< [\prod_{i=1}^{n} a_i , \prod_{i=1}^{n} b_i , \prod_{i=1}^{n} c_i ; Min(\mu_{11}, \mu_{21}, ..., \mu_{n1}); Min(\mu_{21}, \mu_{22}, ..., \mu_{n2})] >$. (5-25)

Figure(5-7). Configuration of a series system

ii. **Parallel system**

For the parallel system shown in Figure(5- 8), where the reliability $\tilde{R}_i$ of component P_i is represented by a vague set

$< [(a_i, b_i, c_i); \mu_{i1}; \mu_{i2}] >$, where $0 \le \mu_{i1} \le \mu_{i2} \le 1$and $1 \le i \le n$. Then, the reliability $\tilde{R}_i$ of the parallel system shown in Figure 6 can be evaluated as follows:

$$\tilde{R} = 1 \ominus \prod_{i=1}^{n}(1 \ominus \tilde{R}_i) \ldots\ldots\ldots\ldots\ldots\ldots (5\text{-}26)$$

$$= 1 \ominus (1 \ominus < [(a_1, b_1, c_1); \mu_{11}; \mu_{12}] >) \otimes (1 \ominus < [(a_2, b_2, c_2); \mu_{21}; \mu_{22}] >) \otimes \ldots \otimes (1 \ominus < [(a_n, b_n, c_n); \mu_{n1}; \mu_{n2}] >)$$

$$= 1 \ominus < [(1-c_1, 1-b_1, 1-a_1); \mu_{11}; \mu_{12}] > \otimes < [(1-c_2, 1-b_2, 1-a_2); \mu_{21}; \mu_{22}] > \otimes \ldots$$

$$\otimes < [(1-c_n, 1-b_n, 1-a_n); \mu_{n1}; \mu_{n2}]$$

$$= 1 \ominus < \left[\begin{matrix} \prod_{i=1}^{n}(1-c_i), \prod_{i=1}^{n}(1-b_i), \prod_{i=1}^{n}(1-a_i); \\ Min(\mu_{11}, \mu_{21}, \ldots, \mu_{n1}); Min(\mu_{21}, \mu_{22}, \ldots, \mu_{n2}) \end{matrix}\right] >$$

$$= < \left[\begin{matrix} (1-\prod_{i=1}^{n}(1-a_i), 1-\prod_{i=1}^{n}(1-b_i), \prod_{i=1}^{n}(1-c_i)); \\ Min(\mu_{11}, \mu_{21}, \ldots, \mu_{n1}); Min(\mu_{21}, \mu_{22}, \ldots, \mu_{n2}) \end{matrix}\right] > (5\text{-}27)$$

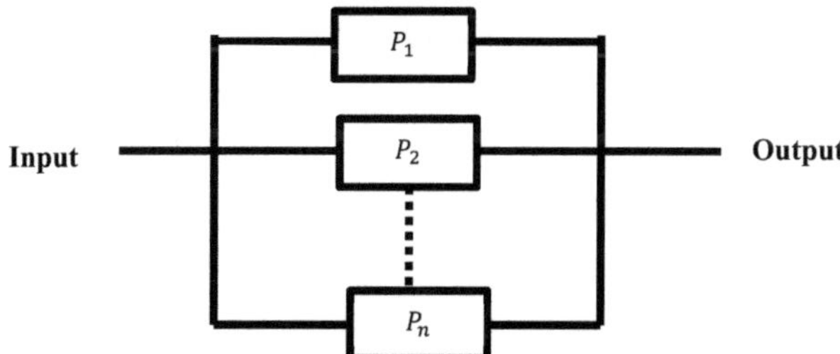

Note that 5.5.4·

Figure(5- 8). Configuration of a parallel

For two triangular vague sets A and B there exists two cases as follows :

Case 1: For the triangular vague sets $\tilde{A}$ and $\tilde{B}$ shown in Figure (5-9), where

$$\tilde{A} = < [(a_1, b_1, c_1); \mu_1], [(a_1, b_1, c_1); \mu_2] > = < [(a_1, b_1, c_1); \mu_1; \mu_2] >$$

$$\tilde{B} = < [(a_2, b_2, c_2); \mu_1], [(a_2, b_2, c_2); \mu_2] > = < [(a_2, b_2, c_2); \mu_1; \mu_2] >$$

and $0 \leq \mu_1 \leq \mu_2 \leq 1$.

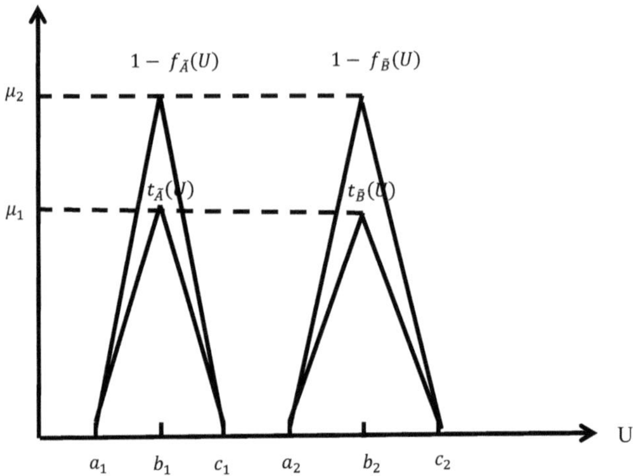

Figure (5-9). Triangular vague sets $\tilde{A}$ and $\tilde{B}$(Case 1)

Case 2: Consider the triangular vague sets $\tilde{A}$ and $\tilde{B}$ shown in Figure (5-10), where

$$\tilde{A} =< [(a_1, b_1, c_1); \mu_1], [(a_1, b_1, c_1); \mu_2] >,$$

$$\tilde{B} =< [(a_2, b_2, c_2); \mu_3], [(a_2, b_2, c_2); \mu_4] >,$$

and $0 \leq \mu_3 \leq \mu_1 \leq \mu_4 \leq \mu_2 \leq 1$

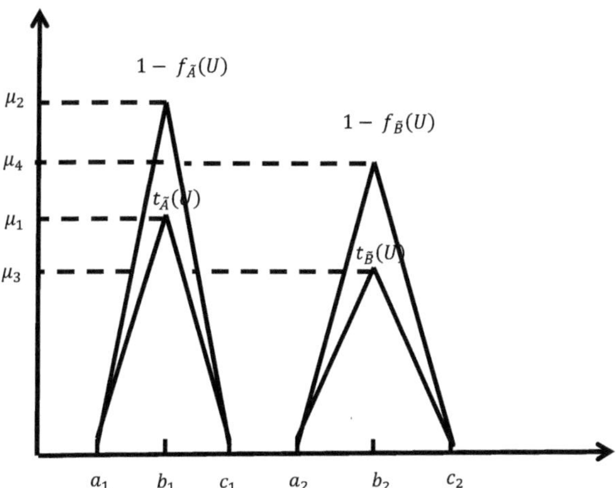

Figure (5-10). Triangular vague sets $\tilde{A}$ and $\tilde{B}$ (Case 2)

It can retain to [70] about the arithmetic operation between A and B

Example 5.5.5: consider the system shown in Figure below, where the reliability of the components C_1, C_2, C_3 , and , C_4 are $\tilde{R}_1, \tilde{R}_2, \tilde{R}_3, and\ \tilde{R}_4$, respectively

$$\tilde{R}_1 = < [(a_1, b_1, c_1);\ \mu_{11}, \mu_{12}] >,$$

$$\tilde{R}_2 = < [(a_2, b_2, c_2);\ \mu_{21}, \mu_{22}] >,$$

$$\tilde{R}_3 = < [(a_3, b_3, c_3);\ \mu_{31}, \mu_{32}] >,$$

$$\tilde{R}_4 = < [(a_4, b_4, c_4);\ \mu_{41}, \mu_{42}] >,$$

$0 \leq \mu_{i1} \leq \mu_{i2} \leq 1$, and $1 \leq i \leq 4$.

C_1 C_3

Input output

C_2 C_4

Figure (5-11) A series – parallel system

The reliability of the system $\tilde{R}_S$ can be evaluated as follows:

$\tilde{R} = [1\ominus (1\ominus\tilde{R}_1) \otimes (1\ominus\tilde{R}_2)] \otimes$
$[1\ominus (1\ominus\tilde{R}_3) \otimes (1\ominus\tilde{R}_4)]$..................................(5-28)
$= [1\ominus (1\ominus< [(a_1, b_1, c_1);\ \mu_{11}\ ;\ \mu_{12}] >) \otimes$
$(1\ominus < [(a_2, b_2, c_2);\ \mu_{21}\ ;\ \mu_{22}] >)] \otimes$
$[1\ominus (1\ominus< [(a_3, b_3, c_3);\ \mu_{31}\ ;\ \mu_{32}] >) \otimes$
$(1\ominus< [(a_4, b_4, c_4);\ \mu_{41}\ ;\ \mu_{42}] >)]$
$= [1\ominus< [(1 - c_1\ , 1 - b_1, 1 - a_1);\ \mu_{11}\ ;\ \mu_{12}] > \otimes$
$<[(1 - c_2, 1 - b_2, 1 - a_2);\ \mu_{21}\ ;\ \mu_{22}] > \otimes$
$[1\ominus< [(1 - c_3, 1 - b_3, 1 - a_3);\ \mu_{31}\ ;\ \mu_{32}] > \otimes$
$< [(1 - c_4, 1 - b_4, 1 - a_4);\ \mu_{41}\ ;\ \mu_{42}] >]$
$= [1\ominus< [((1 - c_1\ )(\ 1 - c_2), (1 - b_1)(1 - b_2), (1 - a_1)$
$(1 - a_2); Min(\mu_{11}\ ;\ \mu_{21}); Min(\mu_{12}\ ;\ \mu_{22})] >] \otimes$
$[1\ominus<[((1\text{-}c_3)(1 - c_4), (1 - b_3)(1 - b_4), (1 - a_3)$
$(1 - a_4)); Min(\mu_{31}\ ;\ \mu_{41}); Min\ (\mu_{32}\ ;\ \mu_{42})] >]$...........(5-29)

5.6 . Evaluation of Fuzzy System Reliability by using Soft Intuitionistic Fuzzy Set (briefly SIFS) for Some Systems of System Reliability

In this section we derive the fuzzy reliability for series and by taking the complement of the formula of series system to obtain the formula of parallel system , moreover using the same method to derive the fuzzy reliability of series-parallel system R_{sp} and for parallel series system $R_{ps}$with examples.

Definition5.6.1:

Let $\tilde{R}_\alpha = (\tilde{R}_{\mu_\alpha}, R_{v_\alpha})$ be an intuitionistic fuzzy subset of $X \times Y$ and $\tilde{R}$ defined as intuitionistic fuzzy relationship from X to Y . We write $X \xrightarrow{\tilde{R}} Y$ $\tilde{R}(x, y)$denoted the degree of correspondence between x and y based on the relationship $\tilde{R}$. Let $F(X \times Y)$ denoted the family of an intuitionistic fuzzy relationship on X to Y the set $\tilde{R}_\alpha = \{(x, y) \in X \times Y; \tilde{R}_{\mu_\alpha}(x, y) \geq \alpha$ and $R_{v_\alpha}(x, y) \leq \alpha\} \subset X \times Y$ defined as $\alpha - cut\ set\ if\ \tilde{R} \in F(X \times Y)$for $\alpha \in [0,1]$.

Definition5.6.2:

Let X be an itial set and E be a set of parameters. Let $\mathrm{P}(X)$ denotes the power set of X. Let $A \subset E$. A pair (F, A) is called a soft intuitionistic fuzzy set over X where F is a mapping given by $F: A \to \mathrm{P}(X)$ and $F(x) = \{y \in \mathrm{X}: (x, y) \in R_{\alpha}, x \in A, y \in X, \alpha \in [0,1]\}$.

Systems of systems 5.6.3:

i. **For series system**

The reliability of every component of series System taken as SIFS for treatment the incompleteness and imprecise in information. The fuzzy reliability (R_s) of series System consists of n component is as follows:

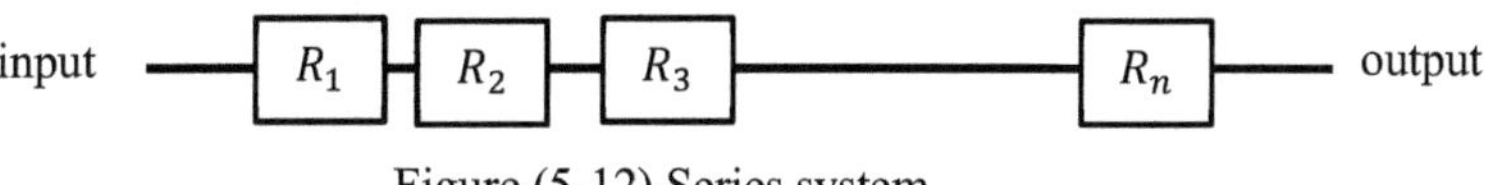

Figure.(5-12) Series system

Let $R = \{R_{i1}, R_{i2}\}$ be the reliability of the ith component taken as SIFS where $R_{i1} = \{\mu_{i1}, v_{i1}\}$ and $R_{i2} = \{\mu_{i2}, v_{i2}\}$, where μ_{i1}, v_{i1} and μ_{i2}, v_{i2} are represents membership and non-membership functions of upper values lower values respectively.

Then the reliability of the series System R_s is :

$$R_s = \prod_{i=1}^{n} R_i = R_1 \times R_2 \times \cdots \times R_n \text{ ....(5-30)}$$

$$= \{(R_{11}, R_{12}) \times (R_{21}, R_{22}) \times \cdots \times (R_{n1}, R_{n2})\}$$

$$= \{(R_{11} \times R_{21} \times \cdots \times R_{n1}), (R_{12} \times R_{22} \times \cdots \times R_{n2})\}$$

$$= \{((\mu_{11}, v_{11}) \times (\mu_{21}, v_{21}) \times \cdots \times (\mu_{n1}, v_{n1})), ((\mu_{12}, v_{12}) \times (\mu_{22}, v_{22}) \times \cdots \times (\mu_{n2}, v_{n2}))\}$$

$$= \Big\{\{u_{11}, u_{21}, \dots, u_{n1}, \textstyle\sum_{i=1}^{n} v_{i1} - \sum_{i,j=1}^{n} v_{i1}v_{j1} - \sum_{i,j,k=1}^{n} v_{i1}v_{j1}v_{k1} \dots + (-1)^{n-1} v_{11}v_{21} \dots v_{n1}\}, \{u_{12}, u_{22}, \dots, u_{n2}, \sum_{i=1}^{n} v_{i2} - \sum_{i,j=1}^{n} v_{i2}v_{j2} - \sum_{i,j,k=1}^{n} v_{i2}v_{j2}v_{k2} \dots + (-1)^{n-1} v_{12}v_{22} \dots v_{n2}\}\Big\} \text{ ....(5-31)}$$

Example5.6.4:

Suppose a series system consists of two components with the reliabilities, by using arithmetic operations in [71,135]

{R } = {( )}

{( )},

{R } = {( )}

{(0.)}.

The reliability of the system is

{( )} × {( )}

{( )}

ii. **Parallel System :**

The reliability of every component of parallel System considered as ISFS for dealing with incompleteness and imprecise in information. To calculate the reliability for parallel System we take the complement for the right equation of series System .So the fuzzy reliability of parallel System consisting of n components is as follows:

If R is the reliability of the ith component.

as show below

(Π) Π ....(5-32)

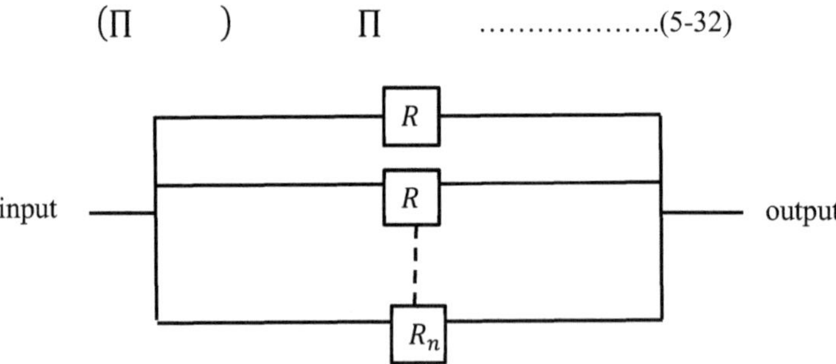

Figure(5-13) Parallel System

(5-33)

And from definition of the complement

{(   )}

$(\mu$) Let − and −

$(1 -$ − $) = ($ $)$

Now

[{(   )},{(

)}]

$[\{p$ Σ $-\Sigma$ $-\Sigma$

$(-1$ $\}, \{p$ Σ $-\Sigma$ $-$

Σ $(-1$ }] ....(5-34)

Therefore

− − − −

Example5.6.5: Suppose the two components connected in parallel, is

$\{(0.7,0.3),(0.5,0.5)\}^c \times \{($ $)\}$

{( )} {( )}

{( )}

{( )}

iii. Series Parallel System.

Consider a system connected in series and each one contains component which are connected in parallel.

If $R_i^j = \left(R_{1i}^j, R_{2i}^j\right)$ is the reliability of the ith component taken as SIFS where is $R_{i1}^j = \left(\mu_{i1}^j, v_{i1}^j\right)$ and $R_{i1}^j = \left(\mu_{i2}^j, v_{i2}^j\right)$. The fuzzy Reliability of the system R_{SP} as below:

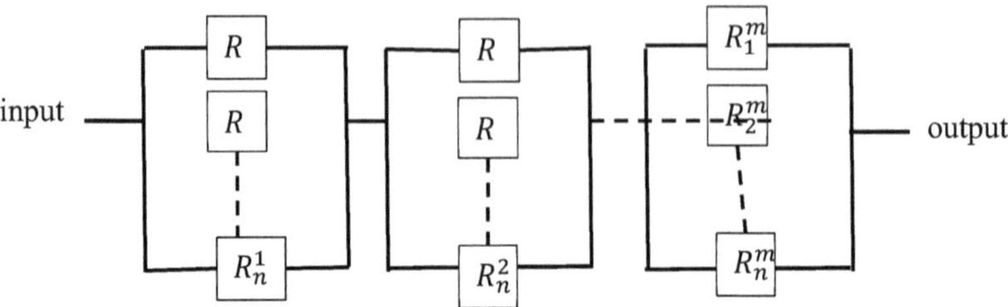

Fig.(5-14) Series – Parallel system

Since the connection appears to be series (form subsystems) and each subsystem connected in parallel. So we can write R_{SP} as follows

$$\prod \left[\prod \quad \right] \prod \left[\prod \quad \prod \quad \right]$$

$$\left([\Pi \quad]\;[\Pi \quad]\right)$$ ....(5-35)

Where

$$\prod$$

$$(\mu \quad) \times (\mu \quad) \qquad (\mu \quad)$$

$$\Big\{\mu \qquad \sum - \sum \qquad - \sum$$

$$+ (-1) \qquad \Big\}.$$

We use the same step for as below

$\prod$ ....(5-36)

$$(\mu \quad) \times (\mu \quad) \qquad (\mu \quad)$$

$$\Big\{\mu \qquad \sum - \sum \qquad - \sum$$

$$(-1 \qquad \Big\}.$$

iv. Parallel Series System

Consider a system connected in series and each one contains component which are connected in parallel and each one contains m component as depicted below:

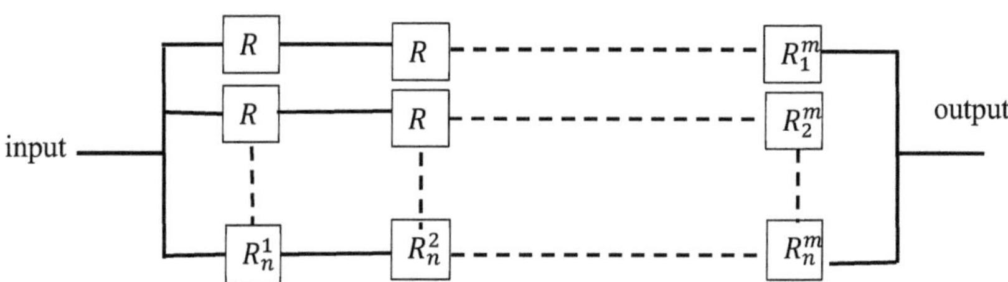

Fig.(5-15) Parallel Series system

If $R_i^j = (R \quad)$ is the reliability of ith component taken as SIFS.
The reliability of the system is as follows:

$\prod_{i=1}^{n} \left(\left[\prod_{j=1}^{m} R_{i2}^{j}\right] [\prod \quad] \right)$, ...(5-37)

$\prod$

$\{\mu$ $,\Sigma$ $-\Sigma$ $-$

Σ $(-1$ $\}$.............(5-38)

And

$\prod$

$\{\mu$ $,\Sigma$ $-\Sigma$ $-\Sigma$

$(-1)^{n-1} v_{i2}^{1} v_{i2}^{2} \dots v_{i2}^{m}\}$. (5-39)

Example5.6.6:

Suppose a series-parallel system consists of two subsystems which are connected in series and each of them contains two components, then the reliability of the system R_{sp} is as follows:

we have $R_{sp} = \prod_{j=1}^{2} \left[\prod_{i=1}^{2} R_i^j\right]^c$ ....(5-40)

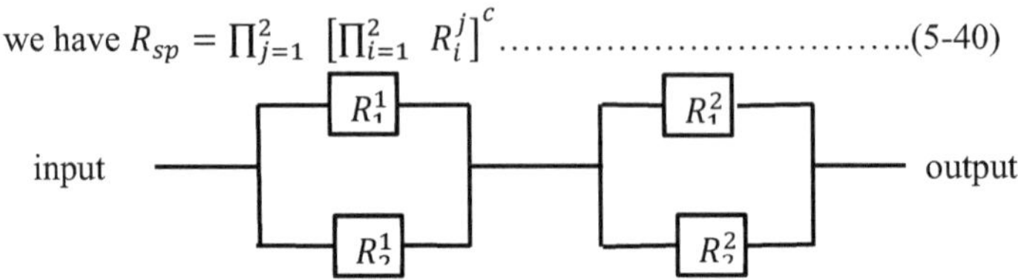

Figure(5-16) Series Parallel system of example 5.6.6

Discussion5.6.7:

1. Any component of a system are represented by soft intuitionistic fuzzy set.
2. To calculate the reliability of a parallel system, we can take the complement for the formula of the series system to get the formula of the reliability R_p for the parallel system
3. The results of examples support the fact that the reliability is minimum in the case of series connection.
4. We can get the same result by using vague sets.

5.7 Appling A triangular Interval Type-2 Fuzzy Set Technique to Calculate the Reliability of Different Models of Systems

Definition

A triangular interval type-2 fuzzy set $\tilde{A} = \langle [(a_1, a_2, a_3), \mu_{\tilde{A}}(x)], [(b_1, b_2, b_3), V_{\tilde{A}}(x)] \rangle$ the membership function is denoted as $\mu_{\tilde{A}}(x)$

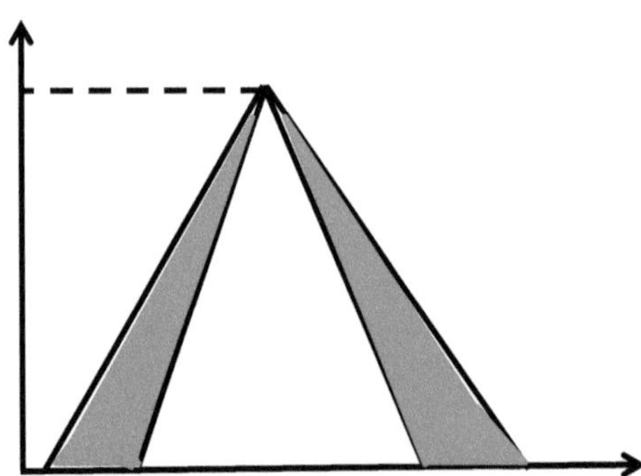

Figure (5- 17). A Triangular Interval type -2 fuzzy membership function

Definition 5.7.2: An interval type-2 number is interval type-2 set whose lower and upper member ship function of it's domain of uncertainty(bravely DOU) are type-1 fuzzy number where $(DOU(\tilde{A}) = \{(x,u) \in X \times [0,1], \mu_{\tilde{A}}(x,u) > 0\}$

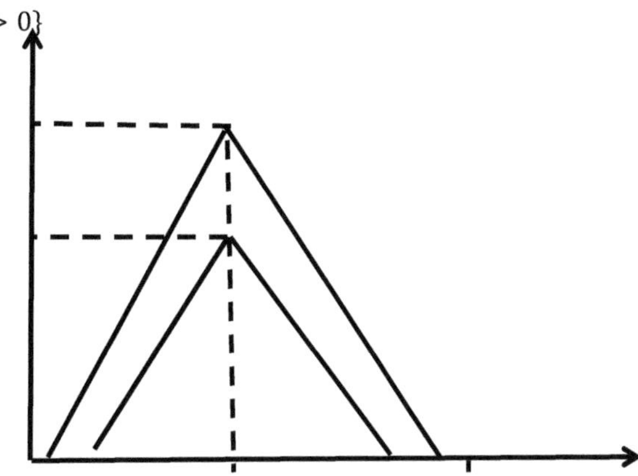

Figure (5- 18). A Triangular interval type -2 fuzzy

Fuzzy Reliability for Different Models of Systems 5.7.3:

We take the reliability of any component to be a triangular interval type-2 fuzzy set and evolve a fuzzy reliability evaluation for mixed system. Let a system consist of n components, the interval type-2 fuzzy $\tilde{R}_j$ sets, $j = 1,2,3,\dots,n$ are taken to resemble the reliability of each component

i. For series system as in the figure below the reliability $\tilde{R}_s^i$ is defined as follows:

Figure (5- 19). A series system

$\tilde{R}_S = \prod_{j=1}^{n} \tilde{R}_j = < [(\prod_{i=1}^{n} a_{1i}, \prod_{i=1}^{n} a_{2i}, \prod_{i=1}^{n} a_{3i}): max_{i=1,\dots,n}\mu_{\tilde{A}}(x)],$

$[(\prod_{i=1}^{n} a_{1i}, \prod_{i=1}^{n} a_{2i}, \prod_{i=1}^{n} a_{3i}): max_{i=1,\dots,n} V_{\tilde{A}}(x)] >$ ....(5-41)

ii. For components connected in parallel, the reliability $\tilde{R}_P$ of the parallel system as shown in the figure below is defined as the following:

iii.

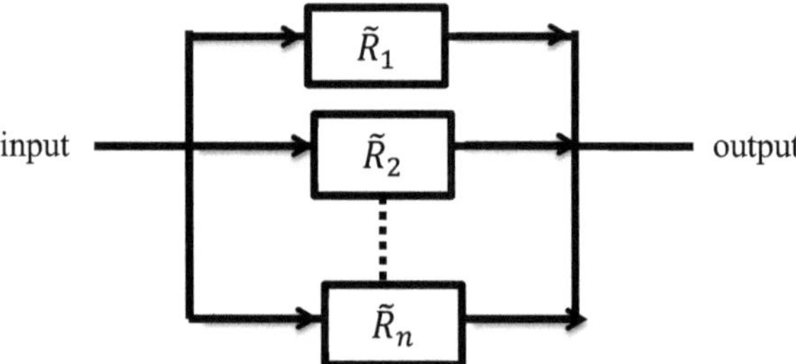

Figure (5-20). A parallel system

$\tilde{R}_P = 1 - \prod_{i=1}^{n}(1 - R_i)$............................(5-42)

$$= \left\langle \begin{matrix} \left[\left(1-\prod_{i=1}^{n}(1-a_{1i}), 1-\prod_{i=1}^{n}(1-a_{2i}), 1-\prod_{i=1}^{n}(1-a_{3i})\right): min_{i=1,\dots,n} M_{\tilde{A}}(x)\right], \\ \left[\left(1-\prod_{i=1}^{n}(1-a_{1i}), 1-\prod_{i=1}^{n}(1-a_{2i}), 1-\prod_{i=1}^{n}(1-a_{3i})\right): max_{i=1,\dots,n} V_{\tilde{A}}(x)\right] \end{matrix} \right\rangle$$

....(5-43)

iii. For mixed system:

a. Consider a system connected in parallel-series consist of m paths connected in parallel, every path contains n components as shown in figure below.

$R_{ps} = 1 - \otimes_{k=1}^{m}(1 - \otimes_{i=1}^{n} R_{ki})$.........................(5-44)

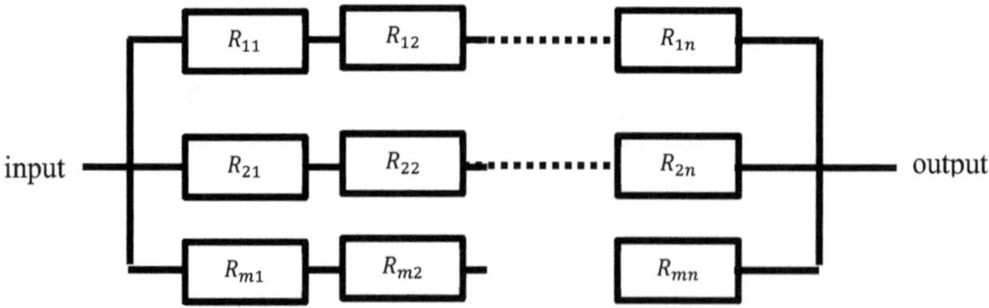

Figure (5-21). Parallel - series system

b. Consider a system connected in a series-parallel consist of n stages connected in series, each one contains m components as shown in the figure below. The fuzzy reliability of the system R_{ps} is:

$$\tilde{R}_{sp} = \prod_{k=1}^{n}\left(1 \ominus \otimes_{i=1}^{m} \left(1 \ominus \tilde{R}_{ik}\right)\right) \text{...........(5-45)}$$

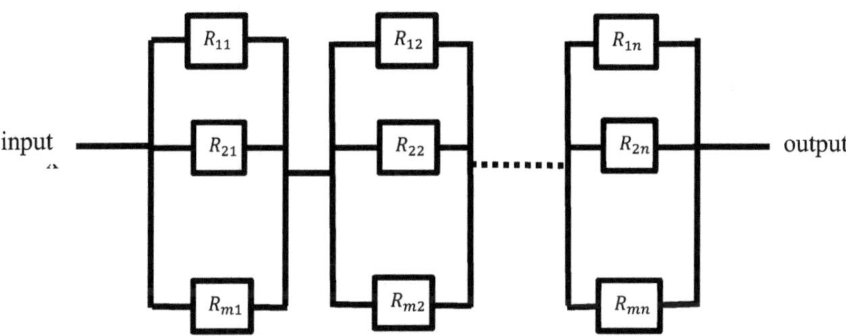

Figure (5-22). Series Parallel system

Example5.7.4:

Calculating the reliability of the system R_{pS} for the bridge system which is depicted in the figure(5-4) , $R_{x_1} = 0.91, R_{x_2} = 0.92$,$R_{x_3} = 0.93, R_{x_4} = 0.94, R_{pS} = 0.9993009651$.We can compare the results for the same example by using different methods.

5.8 Fault Tree Techniques For Analyzing System Reliability

Definition 5.8.1:

The fault tree is a graphic System of the various parallel and sequential combinations of the fault that can lead to the occurrence of the predefined event or top event.

Definition 5.8.2 :

A fault tree is a graphical representation of certain relations which traces a system hazard back words to each for all its possible causes .

Definition 5.8.3:

A path set is a group of fault tree initiators which if none of them occurs will guarantee that the top event can not occur .

Definition 5.8.4:

A cut set is set of components such that when the components in the cut are removed the system there is no path from one terminal to the other .

Definition 5.8.5 :

A minimal cut set is a least group of fault tree initiators which if all occur , will cause the top event to occur .

Definition 5.8.6:

i. A basic event is a basic initiating fault event required no more development . It is symbol by circle

ii. because it is of insufficient consequence or because information is unavailable. It is symbol by rhom

iii. An Intermediate event is a fault event which occurs because of one or more antecedent causes acting through logic gate, all intermediate events are symbols by rectangle

Definition 5.8.7:

A fault is an abnormal , undesirable state of system or a system element induced by presence of an improper command or absence of proper one , or by a failure .

Definition 5.8.8:

A failure is the termination of the device to perform its required function .

Definition 5.8.9:

A fault tree diagrams are logic block diagrams that display the state of a system (Top event) in terms of the states of its components (basic event) , and they are a graphical design technique .

Definition 5.8.10:

A graph is a set of nodes together with a set of branches with the condition that each branch terminates at each end into a node .

Definition 5.8.11:

A graph G is said to be connected if every two of its vertices are connected . A vertex u is said to be a u – v path in G . i.e., A graph G is connected if it cannot expressed as the union of two graphs , otherwise it is disconnected

Definition 5.8.12:

A walk in a graph G is a finite sequence of edges in which any two consecutive edges are adjacent or identical.

Definition 5.8.13 :

A walk in which all the edges are distinct is a trail. If in addition the vertices distinct except the two ends then the trial is a path.

Definition 5.8.14 :

A connected graph G is called Eulerian if there exist a closed trail containing every edge of G such a trial is an eulerian trail. The definition require each edge to be traversed once and once only.

Logic Gates and Boolean algebra rules 5.8.15:

They are two basic types of fault tree gates , the OR – gate and the AND – gate and all other gates are special cases of these two basic types .

i. The AND Gate

The AND – gate is used to show that the output fault occurs only if all the input fault occur . There may be any number of input faults to an AND – gate. The figure below shows a typical two – input AND – gate with input events Input A and Input B , and output event Q. Output occurs only in Input A and Input B both occur. Whicl mbolized by

ii. The OR Gate

The OR – gate used to show that output event occurs only if one or more of the input events occur . There may be any number of input event to an OR – gate. Which symbolize

For most fault trees, particularly those with one or more primary failures occurring in more than one branch of fault tree, the rules of Boolean algebra may be used to simplified expression for top event T, as absorption , idem potency and De Morgan roles with illustrative examples :

Example 5.8.16:

For the fault tree in figure below

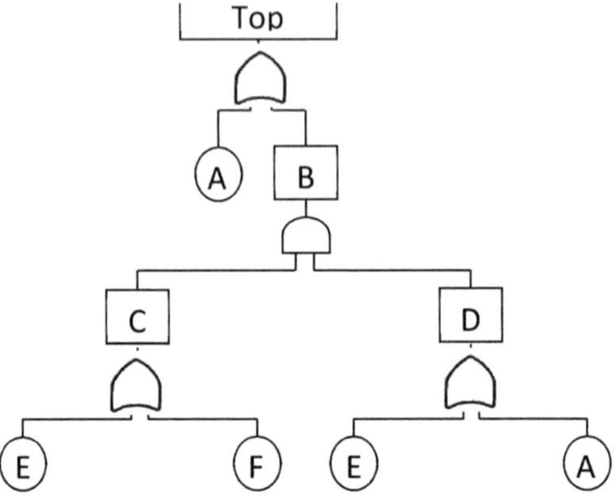

Fig(5-23): Fault Tree of Example 5.6.4

$$
\begin{aligned}
T &= A \cup B \\
&= A \cup [C \cap D] \\
&= A \cup [(E \cup F) \cap (E \cup A)] \\
&= A \cup [\, E \cup (F \cap A)\,] \\
&= (A \cup E) \cup (F \cap A) \\
&= E \cup [\, A \cup (F \cap A)\,] = E \cup A
\end{aligned}
$$

For the fault tree in figure above , the top event will occur if either event A occurs or the event E occurs , the equivalent fault tree(reduced) is shown in the figure below:

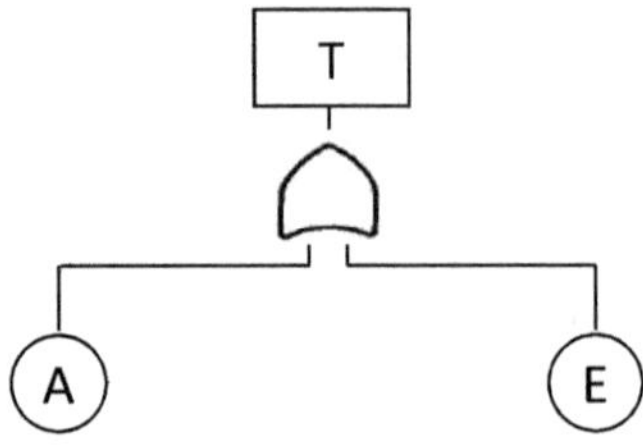

Figure(5-24): Reduced Fault Tree of the Fault tree in figure(5-23).

Example 5.8.17:

The failure in the nuclear reactions because of long time running especially the cooling system and temperature control systems which are very important in the nuclear reactions because of the dangerous of disaster of the explosion as the explosion in Cheronble and Fukushima reactions . Consider the system of jacket water cooling in nuclear reactions as shown in the fault tree below which is taking as problem from [134]

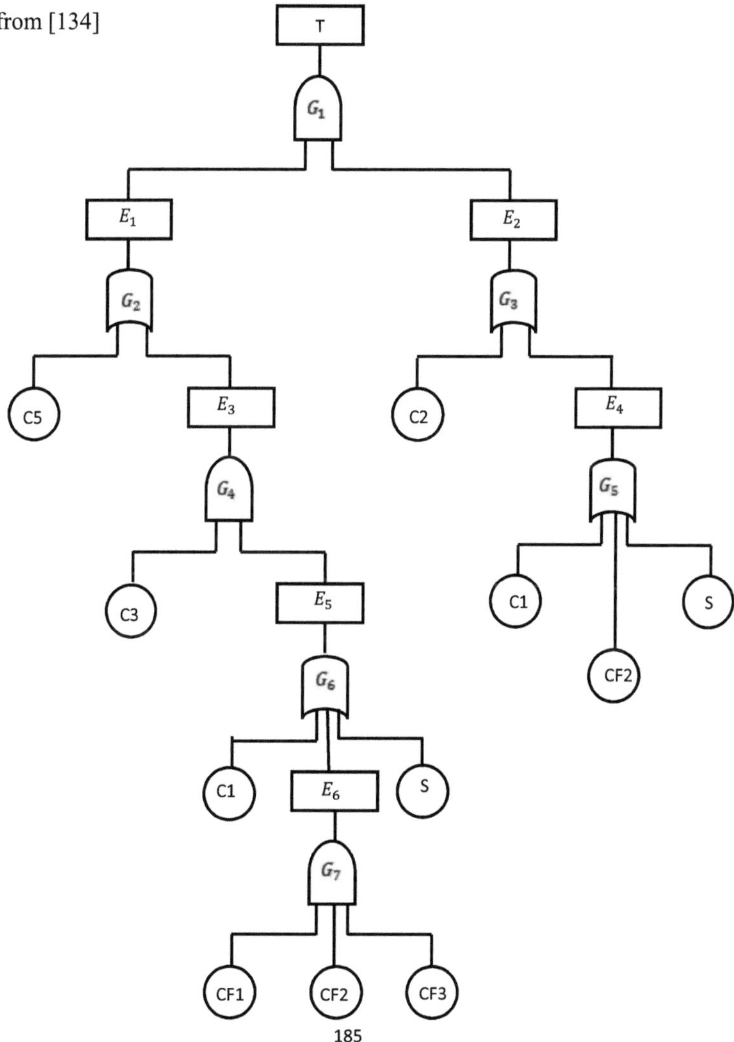

Fig(5-25): A fault tree of power plant system

Where $G_1, G_2, \dots, G_7$ are generators

$C_1, C_2, \dots, C_5$ are cooling units

CF_1, CF_2, and, CF_3 cooling fuel units

And S standby cooling unit

From fault tree above by using the Boolean algebra operations to determine minimal cut sets as follows:-

The top event T=$E_1 . E_2$

$E_1 = C_5 + E_3$

$E_2 = C_2 + E_4$

$E_3 = C_3 . E_5$

$E_4 = C_1 + CF_2 + S$

$E_5 = C_1 + E_6 + S$

$E_6 = CF_1 . CF_2 . CF_3$

Now by substitution

$$T = (C_5 + E_3)(C_2 + E_4)$$

$$= (C_5 + C_3 . E_5)(C_2 + C_1 + CF_2 + S)$$

$$= [(C_5 + C_3(C_1 + S + E_6)](C_2 + C_1 + CF_2 + S)$$

$$= [C_5 + C_3(C_1 (CF_1 . CF_2 . CF_3) + S)] (C_2 + C_1 + CF_2 + S)$$

By using Boolean roles

$$T = [C_5 + (C_3 . C_1) + C_3 . [CF_1 . CF_2 . CF_3) + (C_3 S)](C_2 + C_1 + CF_2 + S)$$

Noted that $C_3 . C_1 . C_1 = C_3 . C_1$

We get the minimal cuts :-

$C_5.C_2,$

$C_5.C_1$

$C_5.CF_2$

$C_5.S$

$C_3.C_1$

$C_3.S$

$C_3.CF_1.CF_2.CF_3$
Noted that any minimal cut sets causes the failure of the system.

Analytical Approaches 5.8.18:

There are two generic analytical methods by means of which conclusion are reached in the human sphere :

i. Inductive Approaches

Induction involves reasoning from individual cases to a general conclusion . If , in the consideration of a certain system , a particular fault or initiating condition is postulated and an attempt to ascertain the effect of that fault or condition on system operation is made , an inductive specified control surface affects the fight of an aircraft or into how the elimination of some item in budget affects the overall operation of a school district . Inductive approaches are also termed bottom – up approaches that the start at the bottom , i.e., at the failure initiators and basic event initiators , and then upwards to determine the resulting system effects of a given initiator . Inductive approaches thus start at a possible basic cause and then analyze the resulting effect.

ii. Deductive Approaches

Deductive constitutes reasoning from the general to the specific . Inductive system analysis , it is postulated that the system itself has failed in a certain way , and an attempt is made to find out what modes of system or subsystem (component) behavior contribute to this failure . All successful detective and other types of investigators are experts in deductive analysis .

Typical of deductive analyses in real life are accident investigation

: what chain of events caused the sinking of an "unsinkable" ship such as the Titanic on its maiden voyage ? What failure processes , instrumental and/or human , contributed loss of the vertical stabilizer and to the crash of commercial airliner into residential area?.

Briefly , inductive approach applied to determine what system states (usually failed states) are possible ; deductive approach which termed top-down are applied to determine how a given system state (usually failed states) can occur .

Example 5.8.19:

Assume we have the following the fault tree which is taken from[161]:

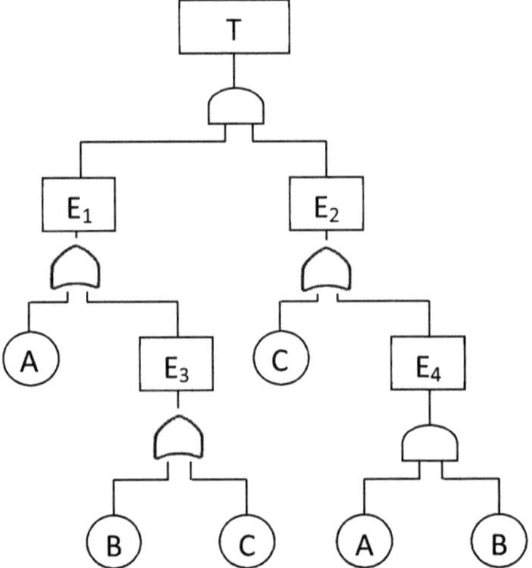

Fig(5.26): Fault tree of a system

The Boolean equations for FT above are :

$T_1 = E_1 \,.\, E_2$

$E_1 = A + E_3$

$E_2 = C + E_4$

$E_3 = B + C$

$E_4 = A \,.\, B$

There are two approaches to solve this system which are :

Frist Approach Bottom – Up (ascending approach):

It begin at the bottom at the tree and pressed up word with the same substitution and expansion technique.

The equation having only basic failure

$E_1 = A + E_3$, $E_3 = B + C$, $E_2 = C + E_4$, $E_4 = A . B$

$E_1 = A + B + C$

$E_2 = C + A . B$

Thus

$T = (A + B + C) . (C + (A.B))$

$= A . C + A (A . B) + B . C + B (A . B) + C . C + C (A . B)$

$= A . C + B . C + C + C (A . B) + A . B + A . B$

$= C + A . B$

We have two minimal cut sets: C, A.B

Second Approach Top – Down (descending approach):

It is start with the top event equation and substitute and expand until getting the minimal cut set expiration for the top event T we have :

$T = E_1 . E_2$

$= [A + E_3] [C + E_4]$

Substitute E_3 by B + C

$= [A + (B + C)] [C + E_4]$

$= A . C + A E_4 + C . E_3 + E_3 . E_4$

$= A . C + A . E_4 + C . B + C . C + B . E_4 + C . E_4$

$= (A . C + C . B + C + C . E_4) + A . E_4 + B . E_4$

$= C + A . E_4 + B . E_4$

$= C + A (A . B) + (A . B)$

$= C + A . B + A . B$

$= C + A . B$

The minimal cut sets are C and A.B

5.9 Event Trees Technique :

A fault tree analysis is used as a tool to perform a system safety analysis and to evaluate reliability of engineering systems.

It is often very difficult to estimate precise rates or probabilities of system component in dynamically changing environments or in system where data insufficient for statistical inference .

The event tree is a quantitative technique for such inductive analysis . It begins with a specific initiating event , a particular cause of an accident , and then follows the possible progressions of the accident according to the success of failure of other components or pieces of equipment . Event Tress are particular adaptation of the more general decision tree formalism that is widely employed for business and economic analysis , they are quite use in analyzing the effects of the functioning or failure of safety systems in response to an accident . Particularly when events follow with the particular time progression .

Suppose that we want to examine the effects of the power failure in a hospital in order to determine the probability of a blackout , a long with other likely consequences . For simplicity we assume that the situations may be analyzed in terms of just three component :

1. The off – side local utility power system that supplies electricity to the hospital .
2. A diesel generator that supplies emergency power .
3. a voltage monitoring system that monitors the off – side power supply and in the event of a failure transmits a signal that starts the diesel generator .

The initiating event is the loss of off – site power . The second is detection of the loss and subsequent functioning of the voltage – monitoring system , and the third event is the start – up and operation of the diesel generator . This sequence is shown in the event tree in figure below. Note that at each event tree there is a branch corresponding to whether a system operates or fails . By convention the upward branches signify successful , and the lower branches failure .

Note that for a sequence of n events there will be 2^n branches of the tree . The number may be reduced by eliminate impossible branches ,

For example the generator cannot start unless the voltage monitor functions . Thus the path is impossible and can be pruned from the tree . as in fig (5-28) voltage diesel off site monitor power

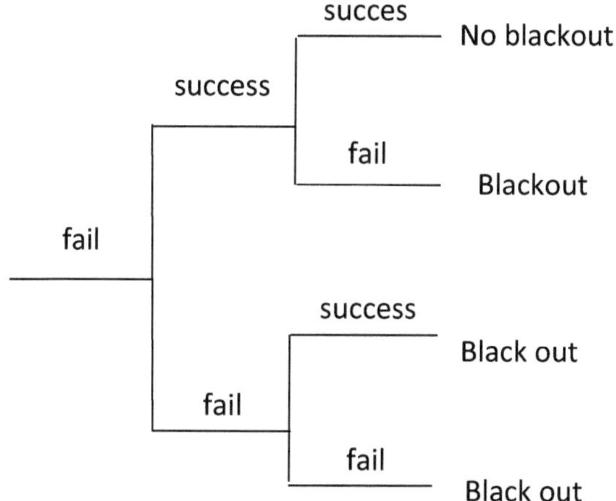

Figure (5- 27) Event tree of Black out

The steps :

1. We may follow an event tree from left to the right to find the probabilities and consequences of differing sequences of events.
2. The probability of the various out comes are determined by attaching a probability to each event on tree . In our tree the probability are p_i for the initial event p_v for the failure of the voltage monitoring and p_g for the failure of the diesel generator . With the assumption that failures are independent , the probability of a blackout is therefore :

$P_i p_v + p_i (1 - p_v) p_g$.

The corresponding fault tree for the example above is as in figure below :

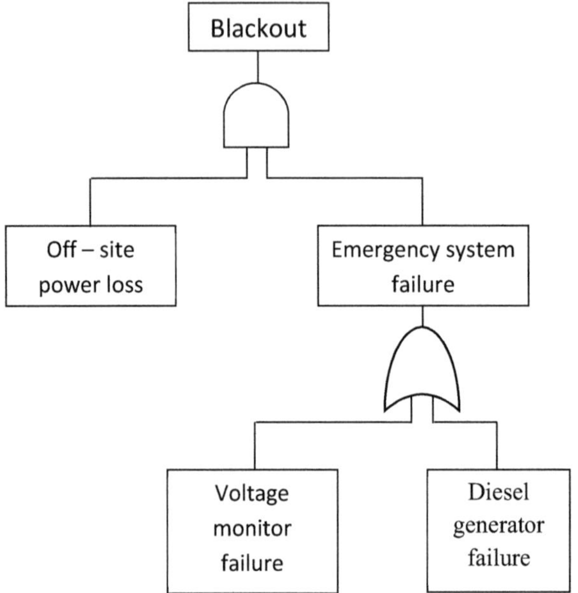

Fig. (5- 28) Fault tree for blackout

Closely associated with fault tree is the event – tree analysis . The analysis involves a sequence of events initiated by a starting or a triggering event . The system fail rely on how important are the subsequent events which are set in motion by the initiating event , therefore we have to catalog all the possible sequences of subsequent events generated by the initiating event . This can be attained systematically and effectively through the use of event tree diagram .

event which gives rise to i number of first subsequence events , these

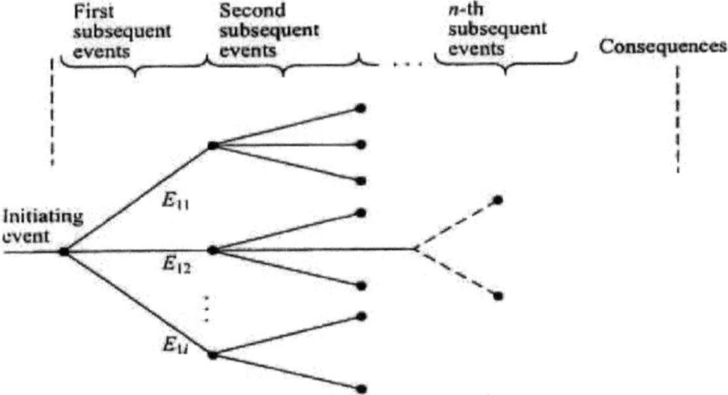

Figure(5-29): A general event-tree model

The general event tree System in figure (5-29)which consists nodes and branch. Each branch stand for a particular event. E is the initiating event which gives rise to i number of first sequence events, these are indicated by E_{11} , E_{12} , … , E_{1i} , Each first subsequence event may act as an initiating event for second subsequent events , and these may be consider as an initiating events for other subsequent event . The ending or the final branches of the System represent the final sequence

Example 5.9.1: The following the example is a modification for an example taken from [131]

Manually operated gates are quite common at railway level crossings . At pre – set times or upon receiving information , the gatekeeper is expected to close the gates on either side of the track . Failure to close the gates, buses carrying passengers trying to cross the tracks, vehicles getting stuck right on the railway tracks , etc… are frequent occurrences.

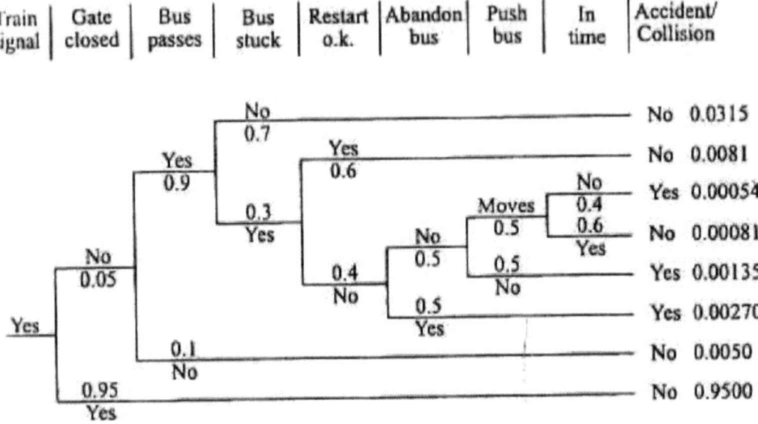

Figure (5-30) show an event – tree model indicating the possible events

along with the event – branches . When the bus gets stuck on the tracks , the driver of the bus , in his anxiety to restart usually drains the battery . Passengers trying to push the bus is another common occurrence . From the final probability values , the probability of an accident/collision is 4.59×10^{-3} .

The relation between Event Trees With Fault Trees 5.9.2:

Each handing of the event tree can be either quantified detailed system Systems to determine the likelihood of system failure , which is oftenly the case of safety systems , or assessed as failure probability , which is mostly the case for the headings of the event tree connected with certain operator actions .

Two general methods or approaches exists for the event tree linking process with the fault tree analysis :

1. Small event tree and large fault tree approach .
2. Large event tree and small fault tree approach . .

The small event tree and large fault tree approach is mostly used in nuclear industries in probabilistic safety assessment of nuclear power plants. The smaller number of safety systems in the event tree heading and more detailed system analyses are performed with the fault tress [89].

Decomposition tree technique 5.9.3

Return to the figure (5-4) which represent a bridge system , the decomposition method depends on choosing one keystone component or more and then applying conditional probability to calculate the system reliability Rs , see[131]

Consider the component A as a keystone , then the system reliability can be expressed as :

$$R_s = P_r\left(S/_A\right)P_r(A) + P_r\left(S/_{A^c}\right)P_r(A^c) \ldots\ldots\ldots\ldots\ldots\ldots\ldots (5-46)$$

Where S = event that the system is successful
A = event the component A is working
A^c= event the component A is failed
$S/_A$ = event or subsystem of that system is working

$S/_{A^c}$ = event or subsystem of that system is failed ,

$$R_s = [1-(1-R_1)(1-R_2)]\,[1-(1-R_4)(1-R_5)].R_3 + [1-(1-R_1R_4)(1-R_2R_5)(1-R_3) \ldots\ldots\ldots\ldots\ldots\ldots.(5\text{-}47)$$

And the decomposition tree which represent the decomposition of the bridge network into two subsystems as shown below:

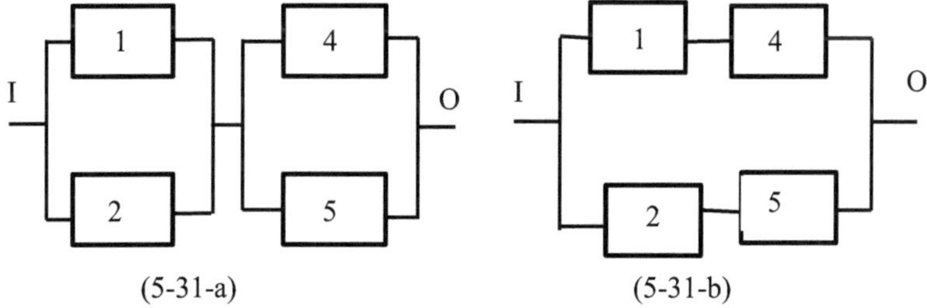

Figure(5-31) , (5-31-a) represent $S/_A$, (5-31-b) represent $S/_{A^c}$ for the system in figure (5-4)

For bridge system in figure (5-4) the decomposition must contain all possible combinations of state for binary components (success, failed) , this means that (2^5) possible combinations of each component which resulting the success or failed of the system as shown in the table below:

Table (5- 13) states for sub systems of figure (5-4)

Component 1	Component 2	component 3	component 4	component 5
S	S	S	S	S
S	S	S	S	F
S	S	S	F	S
S	S	S	F	F
S	S	F	S	S
S	S	F	S	F
S	S	F	F	S
S	S	F	F	F
S	F	S	S	S
S	F	S	S	F
S	F	S	F	S
S	F	S	F	F
S	F	F	S	S
S	F	F	S	F
S	F	F	F	S
S	F	F	F	F
F	S	S	S	S
F	S	S	S	F
F	S	S	F	S
F	S	S	F	F
F	S	F	S	S
F	S	F	S	F
F	S	F	F	S
F	S	F	F	F
F	F	S	S	S
F	F	S	S	F
F	F	S	F	S
F	F	S	F	F
F	F	F	S	S
F	F	F	S	F
F	F	F	F	S
F	F	F	F	F

5.10 Reliability Block Diagram (RBD) and Dynamic Reliability Block Diagram (DRBD):

In this section passive and active faults , RBD and DRBD are presented . A comparison between SFT with RBD, also DFT with DRBD are discussed . Moreover a method to convert DFT into SFT with example is appended .

Passive (static) and active(dynamic) faults (5.10.1:

It can designate the components into their passive or active components according to their function . Pipes , cables bearing , melds and bolts which consider as static components , they function in a more less static manner , often acting as transmitters of energy such as a buss bar or cable , or of fluids such as piping . A static component may usually be thought of as a mechanism for transmitting the output of one dynamic component to the input of another

Active components contribute to system function in dynamic manner , altering in some may the system's behavior for characterizing between static and dynamic components is that the failure rate are normally much higher and dynamic components for static components . Often by two or three orders of magnitude .

The terms active and passive refer to the primary function of the component . Dynamic component may have many static that are prone to failure . for more detail see [134]

Reliability Block Diagram (RBD) 5.10.2:

In a RBD the logic diagram is arranged to indicate which combinations of component failure result in the failure of the system , or which combinations of properly working components keep the system operational . A block in RBD represents the working physical component , and the failure of this component is indicated by the removed of the corresponding block .If enough blocks are removed in an RBD to interrupt the connection between the input and output points , the system is still operating properly . Generally two main types of connection, series and parallel can be established two or more components. The blocks in either series or parallel structure can be merged into a new block . Using such combinations, any parallel-series system can be eventually merged to one block and its reliability can be easily computed by repeatedly using series and parallel structure equations.

Dynamic Reliability Block Diagram (DRBD)5.10.3:

In DRBD system the condition of each component is characterized by a variable state identifying the operational condition of the component at a given time . The evolution of a component's state is a characterized by events occurring to it . The states a generic DRBD component can assume are active if the component works without any problem , failed if component is not operational

RBD VERSUS SFT 5.10.4:

In RBD the system is represented by components connected according to their function or reliability relationship .SFT show which combinations of the components failures causes the system failure .

Note5.10.5 : RBD and SFT Systems are equivalent but DRBD and SFT does not hold because of the following two reasons :

1. It is related to the systeming approach: while SFT and DFT represent only components failure , RBD and DRBD can be also system representations or activations. This fact has no relevance in the static case where the system dynamics is not taken into consideration and activation , thus SFT and RBD are reputed equivalent . But, if dynamics are considered, the possibility of representing reparation and activation events becomes interesting. In fact DRBD can system dynamic behaviors (related to reparations or activations event), DFT is cannot.
2. The difference between DFT and DRBD is the approach to systeming dynamic reliability behaviors: DFT define specific gates to represent failure dependence and/or common mode failure (FDFP), redundancy (spare gate) and order relationships (SEQ and PAND gates), while DRBD explicit and formalize the concept of as the basic to system any dynamic aspect. By this , the DRBD approach offers a degree of freedom in reliability systeming, allowing to represent several dynamic behavior with customizable policies and schemas.

FROM DFT to DRBD 5.10.6:

The mapping from DFT to the DRBD domain is defined. This mapping could be reversed if and only if the DRBD structures exactly coincide with ones specified below, otherwise the DFT analogous of the DRBD structure could not exist .

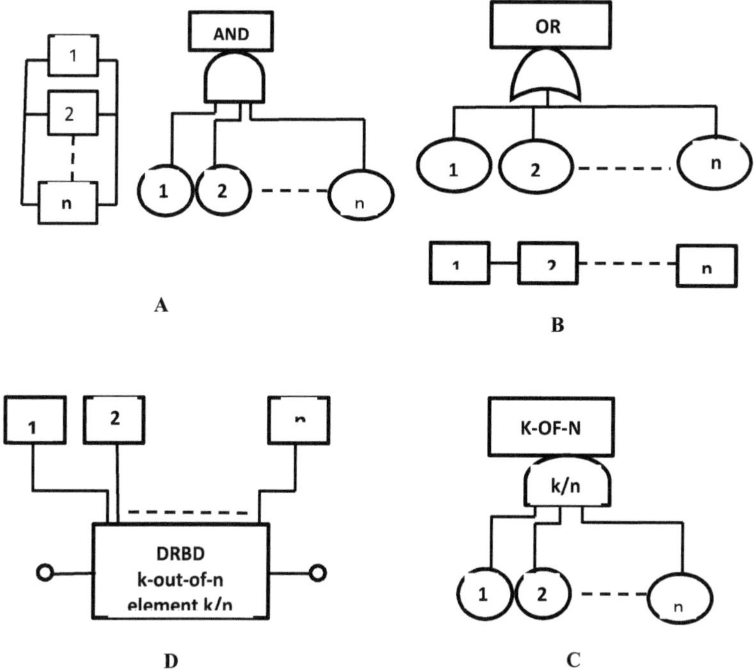

fig(5-32) Comparison between SFT and DRBD,
where A,B and C are SFT while D is DRBD

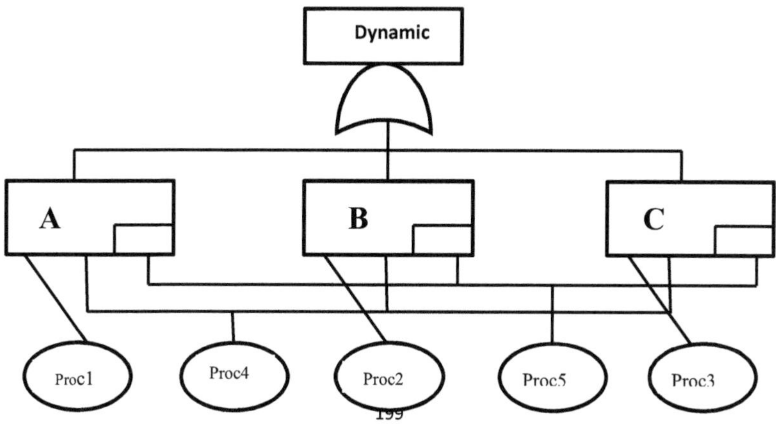

Fig (5-33) Dynamic FT

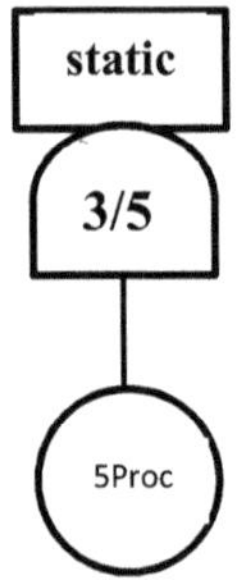

Fig(5-34) static FT

For a static FT in figure above we have ten paths since the combination $C_3^5 = 10$ which are represented as parallel series system.

It is easy to check that the DFT in figure (5-33) and the SFT in figure (5-34) give the same numerical result

Converting dynamic fault tree into static fault tree 5.10.7:

The qualitative analysis consist of computing the failure probabilities of the top event of the SFT , Dugan et. al proposed a new system permeating to include different kinds of temporal and statistical dependencies in the SFT system which called dynamic fault tree (DFT). It is based on definitions of gates which are listed in table (5-14)

Table (5-14). Dynamic gates

Dynamic gate	Notes about input event	Criterion of the failure	Gate figure
Spare gate	Spare gate has one main input and several spare entrances.	When failure of main component happens, first alternative component starts working , after failure of all inputs, gate will have output.	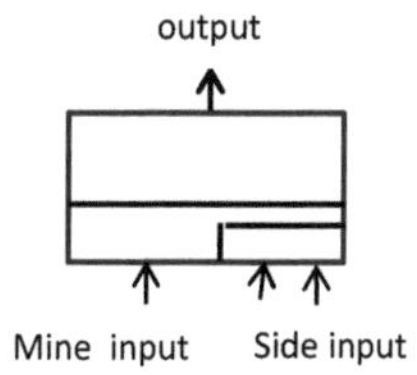
Functional dependency gate FDEP	consists of trigger event and dependent events	If trigger event happens, all dependent event will be happened, and gate will have output .	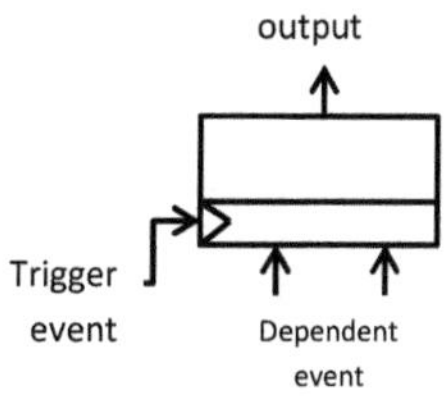
Priority gate (PAND)	Gate has two inputs. A and B, which can be basic events or outputs of other logical gates.	Gate has output if both events A and B, happened, A before B.	
Sequent enforcing gate SEQ	Sequent gate has more inputs.	Gate has output if all inputs in agreed turn.	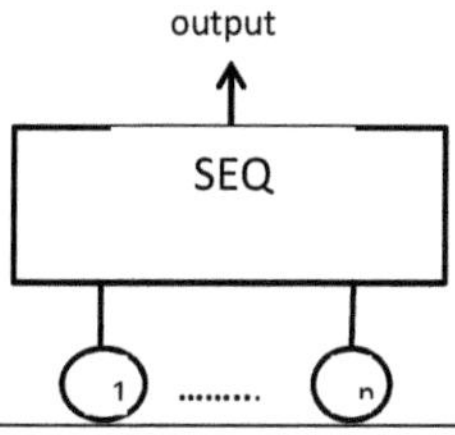

Now in our study a DFT which is taken from [134] is converted into equivalent SFT by replacing a set of Or gates one for each basic event instead of the dynamic gate FDEP to simply the analysis of the reliability for the DFT since it easier for technologists and engineer as illustrated in figures below:

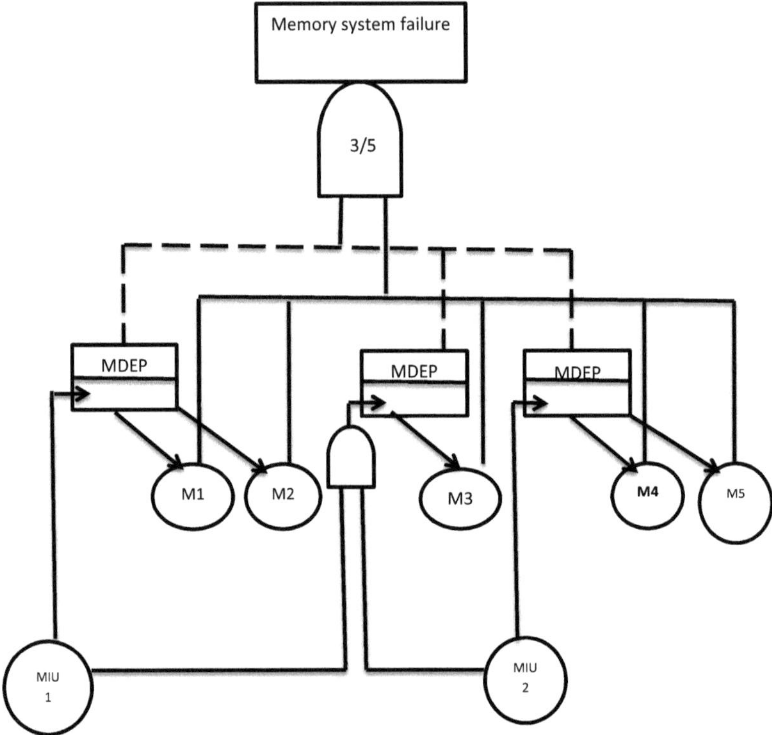

Figure (5-35) DFT which is taken from [134]

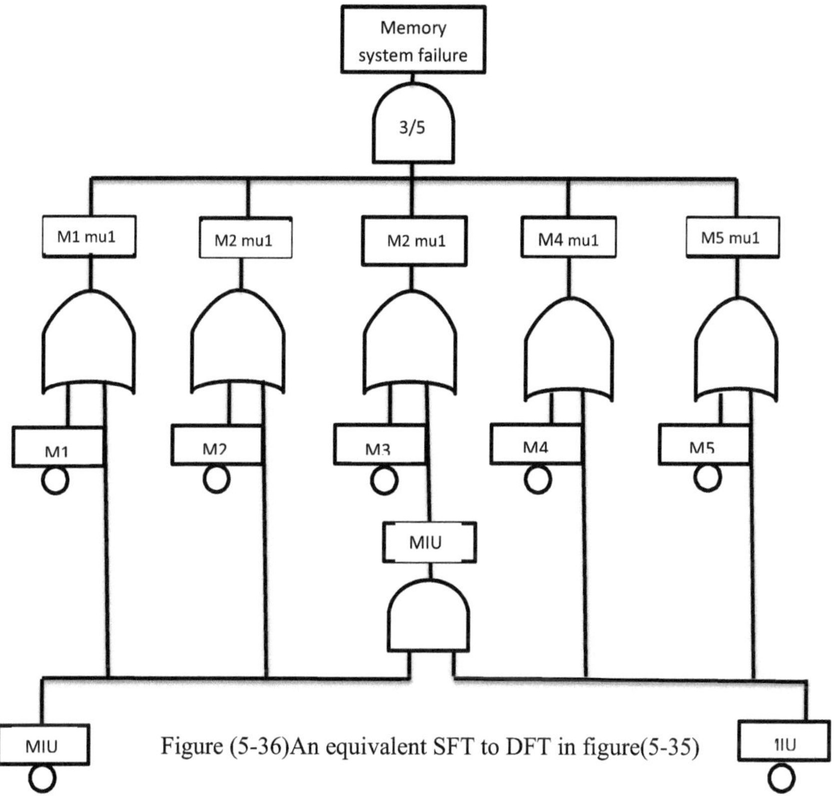

Figure (5-36)An equivalent SFT to DFT in figure(5-35)

5.11 Fuzzy Probist Fault Tree Analysis:

In this section fuzzy probist FTA which is related to the concepts which are discussed in section 5.4 are presented.

Fuzzy Probist Fault Tree 5.11.1:

By Probist fault tree analysis we mean fault tree analysis in the context probist (conventional) reliability theory. Fuzzy probist fault tree analysis applies when the probabilities of basic event in a probist fault tree are represented by fuzzy sets or fuzzy numbers.

Let us consider a probist fault tree depicted in Figure below . We have

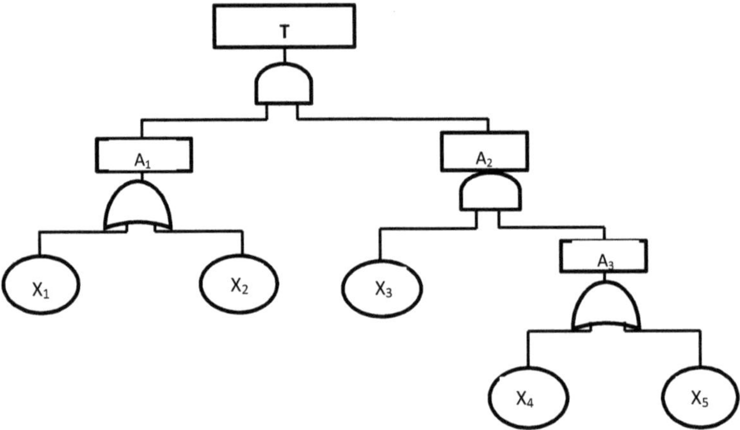

Figure (5-37) An example probist fault tree

– – – –)……..(5-48)

Where we assume basic events $X_1, \ldots, X_5$ are independent , P_T is the failure probability of top event T and P_i the failure probability of X_i . Suppose P_i is a fuzzy number with membership function …,5. Applying the extension principle, P_i is a fuzzy set whose membership function is

$$\mu_T(y) = \begin{cases} \sup_{P_1,\dots,P_5} \min\left(\mu_{x_1}(x_1),\dots,\mu_{x_5}(x_5)\right) & if\ P_T^{-1}(y) \neq \emptyset \\ y = P_T(P_1,\dots,P_5) & \\ 0 & if\ P_T^{-1}(y) = \emptyset \end{cases} \quad \dots \quad (5\text{-}49)$$

Probist Multistate Fault Trees Analysis 5.11.2:

In reliability engineering it is common observation that the systems or components may manifest multistate, as example short circuit fault , open circuit fault and working normally represent capacity multistate. . Consequently , in the corresponding multistate fault tree should constraints, besides that probabilities must take value in [0,1] , imposed on the probabilities of basic events . This represent the main difference between binary-state FTA and multistate FTA. However we note that multistate component on multistate fault trees does not violate the binary-state assumption : the inherent failure criteria are precise or the system component failure have clear-cut definition. so fault tree analysis with multistate components still falls into the scope of probist reliability theory.

Example 5.11.3:

Let us consider FT in figure below , to show that how to do fuzzy probist fault tree analysis with multistate components for a capacitor above . The notation is as follows:

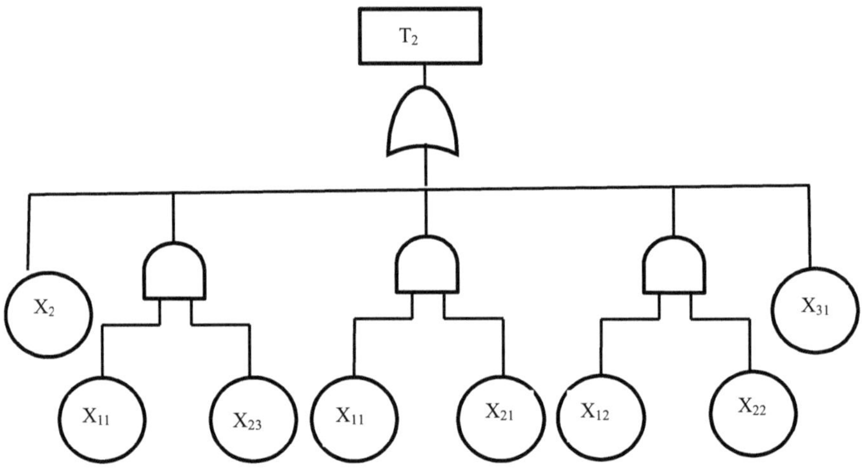

a. Fault Tree for the top event T_2

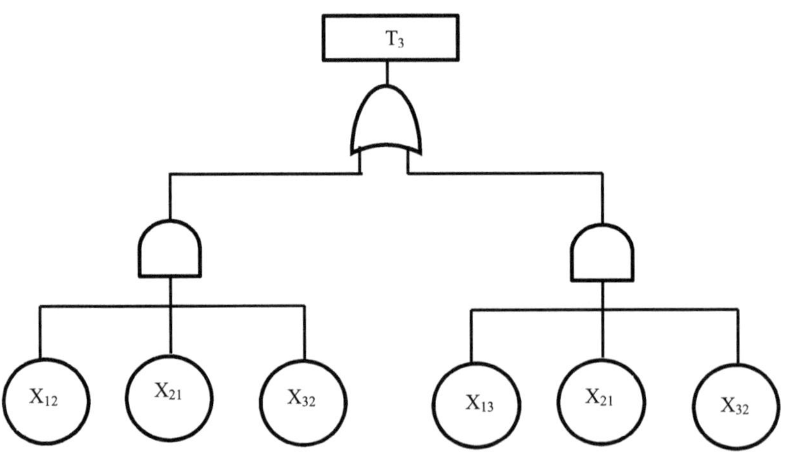

b . Fault Tree for the top event T_3

Figure (5-38) An example5.11.2 of multistate fault tree

X_{11}: short circuit fault

X_{12}: open circuit fault

X_{13}: working normally

X_{21}: short circuit fault

X_{22}: open circuit fault

X_{23}: working normally

X_{31}: open circuit fault

X_{32}: working normally

The corresponding system has three states

T_1: working normally

T_2 : false but safe output signal

T_3 : false and dangerous output signal

Then we have

$T_2 = X_{22} + X_{31} + X_{11}X_{23} + X_{11}X_{21} + X_{12}X_{23}$

$= X_{22} + X_{31}\bar{X}_{22} + X_{12}X_{23}\bar{X}_{31} + X_{11}X_{21}\bar{X}_{31}$

$T_3 = X_{12}X_{21}X_{32} + X_{13}X_{21}X_{32}$ or

$P_{T_2} = P_{22} + P_{31}(1 - P_{22}) + (1 - P_{31})(P_{12}P_{23} + P_{11}P_{23} + P_{11}P_{21})$

$P_{T_3} = P_{12}P_{21}P_{32} + P_{13}P_{21}P_{32}$

Where we assume component states are independent , $P_{T_i}$is the probability of T_i and $P_{ij}$is the probability of X_{ij}. Obviously , there should be

$$P_{11} + P_{12} + P_{13} = 1$$

$$P_{21} + P_{22} + P_{23} = 1$$

$$P_{31} + P_{32} = 1$$

5.12Using Intuitionistic Fuzzy Fault Tree Analysis for Gas Engine Safety System

We use triangular Intuitionistic Fuzzy number to Fault Tree Analysis of gas engine safety system as an application

Definition 5.12.1:

Let a set X be fixed . An intuitionistic fuzzy set(shortly IFS) $\tilde{A}$ in X is an object having the form

$\tilde{A} = \{< x, \mu_{\tilde{A}}(x), v_{\tilde{A}}(x) >: x \in X\}$

$Where\ \mu_{\tilde{A}}(x): X \to [0,1]\ and\ v_{\tilde{A}}(x): X \to [0,1]$ represent the degree of membership and the degree of non-membership respectively of the element $x \in X$ to the set$\tilde{A}, \tilde{A} \subseteq X, \forall x \in X, 0 \leq \mu_{\tilde{A}}(x) + v_{\tilde{A}}(x) \leq 1$

Definition 5.12.2 :

An intuitionistic fuzzy number(briefly IFN) $\tilde{A}$ is defined as follows :

i. Intuitionistic fuzzy subset of the real line
ii. Normal : there is any $x_\circ \in R \ni \mu_{\tilde{A}}\ (x_\circ) = 1\ , v_{\tilde{A}}\ (x_\circ) = 0$
iii. The membership function $\mu_{\tilde{A}}\ (x_\circ)$ is convex
iv. The non- membership function $v_{\tilde{A}}\ (x_\circ)$ is concave

Definition 5.12.3:

Triangular Intuitionistic fuzzy numbers (TIFN):

The TIFN $\tilde{A}$ is an intuitionistic fuzzy set in R with five real numbers $(a_1, a_2, a_3, a', a^{"})$ with $(a' \leq a_1 \leq a_2 \leq a_3 \leq a^{"})$ and the two triangular functions:

$$\mu_{\tilde{A}}\ (x) = \begin{cases} \frac{x-a_1}{a_2-a_1} & a_1 \leq x \leq a_2 \\ \frac{a_3-x}{a_3-a_2} & a_2 \leq x \leq a_3 \\ 0 & elsewhere \end{cases} \quad ....(5\text{-}50)$$

And

$$v_{\tilde{A}}\ (x) = \begin{cases} \frac{a_2-x}{a_2-a'} & a' \leq x \leq a_2 \\ \frac{x-a_2}{a^{"}-a_2} & a_2 \leq x \leq a^{"} \\ 0 & elsewhere \end{cases} \quad ....(5\text{-}51)$$

Intuitionistic Fuzzy arithmetic has the ability to dealing with such cases .

In quantitative evaluation of intuitionistic fuzzy fault tree , these fuzzy data for the individual components are inserted into the tree at the lowest_hierarchical level and combined together using the logic of tree with the intuitionistic Fuzzy arithmetic to give the failure assessment of the complete system being studied .

The fault tree of the gas engine safety system which is shown in figure below is taken for analysis. The failure of gas engine safety system depends on different factors , the tank seep detection failure , heat detection failure , injector system failure

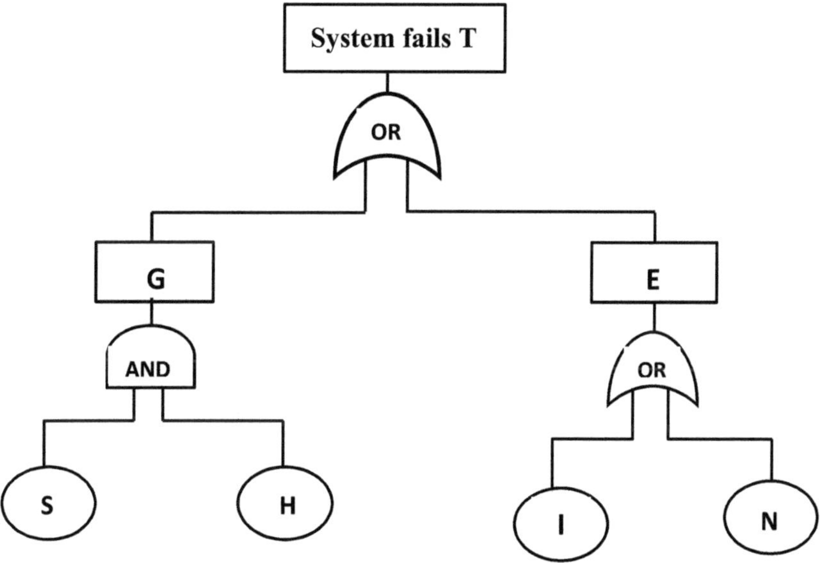

Figure (5-39)Fault tree of failure of gas engine

The system failure (top event) T is:

$T = G \cup E$

Where $G = S \cap H$ and $E = I \cup N$, $T = (S \cap H) \cup (I \cup N)$

Two major factors are:

The gas tank detection failure and the car engine system failure each of them has two-sub factors ,the intuitionistic fuzzy failure of the system can be calculated when the failures of the basic fault events occurrence are known , the failure of the safety system is evaluated as follows :

1. $\tilde{f}_{GS} = 1 \ominus (1 \ominus \tilde{f}_G)(1 \ominus \tilde{f}_E)$

2. $\tilde{f}_G = \tilde{f}_S \otimes \tilde{f}_H$, *and* $\tilde{f}_E = 1 \ominus (1 \ominus \tilde{f}_I)(1 \ominus \tilde{f}_N)$ and the reliability of the system is $R_s = 1 - F_s$
Where:
$\tilde{f}_{GS}$ represent the failure of gas engine safety system
$\tilde{f}_G$ represent the failure of gas detection system .
$\tilde{f}_E$ represent the failure of car gas engine system
$\tilde{f}_S$ represent the failure of seep detection
$\tilde{f}_H$ represent the failure of heat engine detection
$\tilde{f}_I$ represent the failure of injector system

5.13 Calculation of fuzzy system reliability for series and parallel systems by converting the statistical confidence interval to a triangular fuzzy number:

We produced the $(1-\beta)100\%$ confidence interval for all $0.01 \leq \beta < 1$. Starting at 0.01 is arbitrary and you could begin at 0.001 or 0.005 etc.

Denoting these confidence intervals as $[\theta_1(\beta), \theta_2(\beta)]$ for $0.01 \leq \beta < 1$.

Now we put place these confidence intervals, one on top of the other, to produce a triangular shaped fuzzy number $\bar{\theta}$ whose α-cuts are the confidence intervals.

We have $\bar{\theta}[\alpha] = [\theta_1(\alpha), \theta_2(\alpha)]$ for $0.01 \leq \alpha \leq 1$. In this way we are using more information in $\bar{\theta}$ than just a point estimate, or just a single interval estimate.

An interval estimate of the parameter is a segment on the real line and the value of the parameter is thought to lie somewhere on that interval . When an interval estimate of parameter is constructed so that it has a certain known probability of including the value of the parameter between its limits , this interval is called a confidence interval .

It can find the average value of a subsystem by using the stastical data , , suppose $\tilde{R}_j = \frac{1}{}\sum$, R_j is a point estimate of thus $\tilde{R}_j$ is used instead of the reliability of .

If R_j (unknown) , j n is a population reliability of . By the point estimation $\tilde{R}_j$ of then the reliability of :

i. Series system

$$\prod \quad \tilde{R} \qquad ......(5\text{-}52)$$

ii. Parallel system

$$-\prod \quad -\tilde{R} \qquad ......(5\text{-}53)$$

Note 5.13.1: The probability distribution for the error between $\tilde{R}_j$ and is unknown . The $(1-\alpha)$ confidence interval P_j is

$$\left(\tilde{R} \qquad \frac{}{\sqrt{n}}\right) ......(5\text{-}54)$$

Where $\frac{\bar{X}-}{\sqrt{n}}$ ......(5-55)

And S represent the standard deviation[150] .

Note5.13.2:

i. If the chosen value is coincident with $\tilde{R}_j$ then the estimate error (briefly .EE) is simply EE equals zero .
ii. If the value deviates from $\tilde{R}_j$ farther from both sides of $\tilde{R}$ then EE is bigger .
iii. If the value position in one of the two end points in relation (5-54) then EE is maximum .

The error into the confidence level:

i. If the error = 0 the confidence level equals 1.

ii. If the value farther from two sides of $\tilde{R}$, the less the confidence level is at two endpoints $\tilde{R}_j$ in (5-54) , the confidence level is minimized .

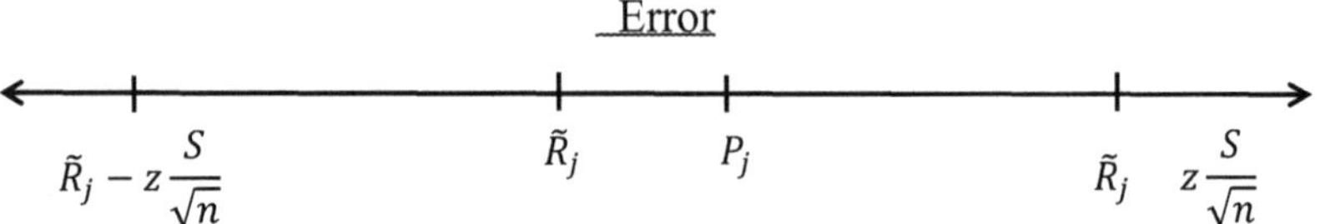

Figure (5-40) The estimate error

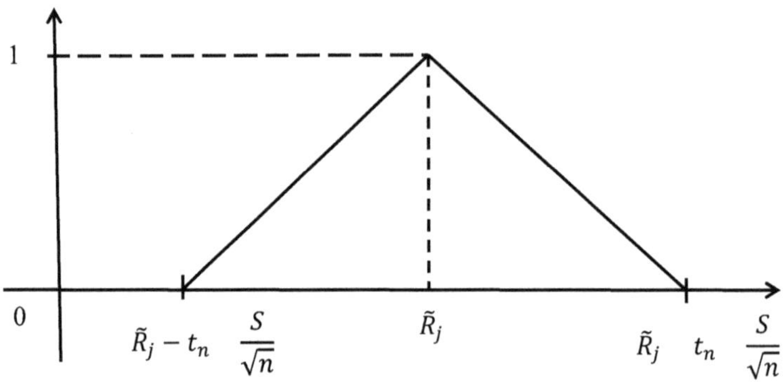

Figure (5-41) $\tilde{R}_j$ triangular fuzzy number

The estimate value of as a fuzzy point , where − then the line of $\tilde{R}_j$,].

Where represent the lower value and be the upper value .

So $\tilde{R}_j -$ − $\sqrt{nj}$ and $\tilde{R}$

− $\left(\overline{\sqrt{nj}}\right)$ ------ (5-56)

Thus $\tilde{R}$]. ------(5-57)

According the relation (5-54) the fuzzy reliability of systems are :

1. For series system : $\tilde{R}_S = \cup\ [\prod_{j=i}^{n} R_{jL}(\alpha), R_{jU}(\alpha) : \alpha]$ -----------(5-58)
2. For parallel system: $\tilde{R}_S = \cup \left[1 - \prod_{j=i}^{n}\left(1 - R_{jU}(\alpha)\right)(\alpha), 1 - \prod_{j=1}^{n} R_{jL}(\alpha) : \alpha\right], j = 1,2, \dots, n$ ……………………………….(5-59)

Example 5.13.3:

Suppose a system consist of two components C_1, C_2, connected in series where

$for\ C_1 : n_1 = 15 , \tilde{R}_1 = 0.89 , S_1 = 0.01$ and $C_2 : n_2 = 20 , \tilde{R}_2 = 0.72 , S_2 = 0.03$

n_i *represnt* the sample size , i = 1, 2

$\tilde{R}_i$ represent the sample mean , i = 1 , 2

S_i represent the standard devian i = 1, 2

$t_{14}(\alpha_1) = 2.592$, $t_{14}(\alpha_2) = 2.694$, $t_{19}(\alpha_1) = 2.512$, $t_{19}(\alpha_2) = 2.694$

The steps of the solution :

1. For C_1 : degree of freedom (df) is = 15 – 1= 14 , and for C_2 = 20 – 1 = 19
2. By sub solution in relation (5-54) we get :
   For $C_1 : [R_{1L}, R_{1U}] = [0.893, 0.916]$ and for $C_2 : [R_{2L}, R_{2U}] = [0.732 , 0.766]$
3. Representing $\tilde{R}_1$ *and* $\tilde{R}_2$ as a triangular numbers as follows :
   $\tilde{R}_1 = (0.893 , 0.910 , 0.916)$ and $\tilde{R}_2 = (0.732 , 0.740 , 0.766)$
4. The $\alpha - level$ fuzzy set of $\tilde{R}_1$ and $\tilde{R}_2$ respectively , $\alpha \in [0,1]$
   Let $A = (a_1, a_2, a_3)$ be a triangular fuzzy number then by $\alpha - cut$ operation interval $A(\alpha)$ is :
   $= [(a_1 - a_2)\alpha + a_1 , -(a_3 - a_2)\alpha + a_3$
   $R_{1L} = (0.910 - 0.893)\alpha + 0.893 = 0.017\alpha + 0.893$
   $R_{1U} = -\ (0.916 - 0.910)\alpha + 0.916\ =\ -0.006\,\alpha + 0.916$, and
   $R_{2L} = (0.740 - 0.732)\alpha + 0.732 = 0.008\,\alpha + 0.732$

$R_{2U} = -(0.766 - 0.740)\alpha + 0.766 = -0.026\,\alpha + 0.766$

5. Calculating $\tilde{R}_S$:

$\tilde{R}_S = \tilde{R}_1 \times \tilde{R}_2 = \cup\, [\prod_{j=1}^{2} R_{jL}(\alpha), \prod_{j=1}^{2} R_{jU}(\alpha): \alpha]$

$$= \bigcup_{0\le\alpha\le1} [(0.017 + 0.893)(0.08\alpha + 0.732)], [(0.916 - 0.006\alpha)(0.766 - 0.02)]$$

$$= \bigcup_{0\le\alpha\le1} [(0.00136\alpha^2 + 0.65367 + 0.01241\alpha + 0.07144), (0.7165 + 0.00012\alpha^2 - 0.023816\alpha - 0.04596\alpha)]$$

$$= \cup_{0\le\alpha\le1}(0.00148\alpha^2 + 0.05545\alpha + 1.35532)$$

6. To estimate the fuzzy reliability $\tilde{R}_S$ via using the signed distance which is defined as follows : see [15]

$$d(\tilde{R}_S, \tilde{0}) = \frac{1}{2}\int_0^1 [\prod_{j=1}^{2} R_{jL}(\alpha), \prod_{j=1}^{2} R_{jU}(\alpha)]\ dx$$

$$= \frac{1}{2}\int_0^1 (0.00148\alpha^2 + 0.05545\alpha + 1.35532)dx$$

$$= \frac{1}{2}[\left(\frac{0.00148\alpha^3}{3} + \frac{0.05545\alpha^2}{2} + 1.35532\alpha\right)]_0^1$$

$$= \frac{1}{2}[\frac{0.00148}{3} + \frac{0.05545}{2} + 1.35532 = \frac{1.383533}{2} = 0.6917665$$

For a system consist of the two components which in example above connected in parallel we apply the same steps according to the formula

$$\tilde{R}_P = \bigcup_{0\le\alpha\le1} [1 - \prod_{j=1}^{n} F_{jL}(\alpha), 1 - \prod_{j=1}^{n} F_{jU}(\alpha): \alpha]\ , where\ F_{jL} = 1 - R_{jL}$$

And $F_{jU} = 1 - R_{jU}$ ....(5-60)

We know that the open intervals consist a base of the usual topology on the real numbers since if G is open sub set of **R** $such\ that. if\ p \in G$ then there exist (a,b) such that $p \in (\mathrm{a}, \mathrm{b}) \subseteq G$

According to above the confidence interval is considered an open interval

Hence $(1-\alpha)100\%$ confidence interval of $\tilde{R}_j$ can be express as $[\tilde{R}_j - \Delta_{jL}, \tilde{R}_j + \Delta_{jU}]$,where the $\Delta_j = tn-1(1-\alpha)\frac{s}{\sqrt{n_j}}$, $0 < \alpha < 1$

For the fuzzy probability $\widetilde{pr}(\tilde{R}_j - \Delta_{jL}, \tilde{R}_j + \Delta_{jU})$

We can say that $\underline{pr}\,(\tilde{R}_j - \Delta_{jL}), \overline{pr}\,(\,\tilde{R}_j + \Delta_{jU})$ represent $pr(int.(\tilde{R}_j - \Delta_{jL}), prcl.(\,\tilde{R}_j + \Delta_{jU})$ respectively

Exercises

1- Write the Boolean expression corresponding to the SFT infigure (5-36) and determine the minimal cut sets .
2- Given an example of fuzzy multi-state Fault tree (FMSFT) to analyze the system reliability.
3- In what situations should we use uncertainty theory?
4- What is meant by :
   a- Ordered fuzzy number
   b- Generalized fuzzy number
5- Given an example of temporal fault tree (TFT) .
6- Combine intuitionistic fuzzy sets and type-2 fuzzy to calculate the reliability of a system in .
7- As there exist a relation between reliability with safety and security ?

8- Apply the decomposition method by choosing two Keystone components to solve example 5.2.7 .

9- For the fault tree in figure (5-39) suppose P(5)= 0.2 , P(H)=0.15 , P(I)= 0.08, P(N) = 0.12 calculate

a- The cut set importance's

b- The component importance's

10-How we can calculate the fuzzy importance of any event in s FT in the

form of fuzzy importance index ?

11-Use soft vague sets to evaluate the reliability of some models and

compare it with the results in section 5.6 .

12-Consider a system in figure below consist of components 1 , 2 and 3

with reliabities R_1 , R_2 and R_3 respectively calculate the system

reliability by using event space method

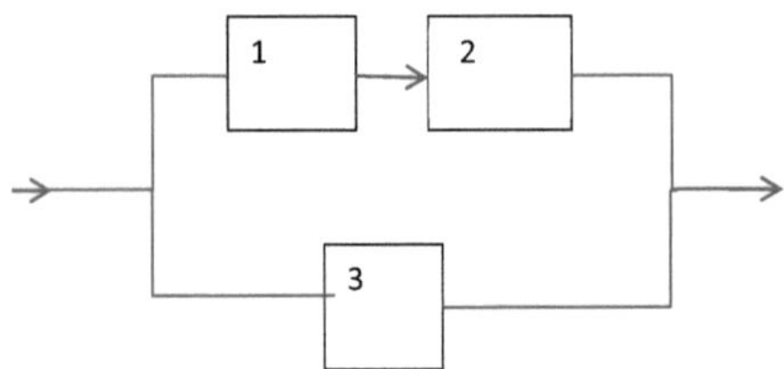

References:

[1]. L.A.Zadeh , "Fuzzy Sets" , Information and Control (1965),pp.,338-353.

[2]. C.L.Chang , "Fuzzy Topological Spacess" , Journal of Mathematical Analysis and Applications 24, ,(1968),pp. 182-190.

[3]. R.Lowen, "Fuzzy Topological Spacess and Fuzzy Compactness" , J.Math. Anal. Appl. 56,(1976), pp. 621-633.

[4]. Didier Dubois ,Henri Prade , "Fuzzy Sets And Systems Theory and Applications",(1980),Academic press.Inc.

[5]. Da-Ruan, "Fuzzy Set Theory and Advanced Mathematical Applications", (1995)Springer Science + Business Media, LLC.

[6]. G. J.Klir , U.S.clair , Bo Yaun - "Fuzzy Set Theory Foundations and Applications" , (1997) ,Prentice Hall PTR.

[7]. G.J.Klir and Bo Yaun , "Fuzzy Logic Theory and Applications" , (2002), Prentice-Hall , 6th edition.

[8]. H.J.Zimmermann , "Fuzzy Set Theory and its Applications", (2001) Science + Business Media LLC .

[9]. G.Chen ,T.T.Pham , "Introduction to Fuzzy Sets , Fuzzy logic , and Fuzzy Control System" , (2001), CRC press LLC.

[10]. R.E.Mercer , J.L.Barron, A.A.Bruen ,D.Chang , "Fuzzy Points: Algebra and Applications", The Journal of Pattern Recognition Society35.(2002) pp. 1153-1166.

[11]. N.palaniaippan "Fuzzy Topology " (2002), Narosa Publications.

[12]. N. Livene."Generalized Closed Sets in Topology " Rend , Circ . Math.Palermo 19(2), (1970),pp. 89-96.

[13]. C.K.Wong "Fuzzy Points Local Properties of Fuzzy Topology "L.Math .Anal.Appl.46,(1974),pp.(316-318).

[14]. H.J.Zimmermann , "Fuzzy Set Theory" , WIREs Computational Statistics volume 2 ,(2010), pp. 317-332.

[15]. M.J.Wierman , "An Introduction to the Mathematics of Uncertainty-program" , (2010),Creighton University.

[16]. Luay A.H.Al-Swidi , and Amed S.A.Oon , "Fuzzy γ-Open and Fuzzy γ-Closed Sets" , American Journal of Scientific research Vol.27 (2011) pp. 63-69.

[17]. B.C.Tripathy and G.C.Ray "Weakly Continuous Functions on Mixed Fuzzy Topological Spacess" Acta Scientiorum , Technology ,v.36,(2014), PP..331-335.
[18]. A.Sadeghiain , J.M. Mendel and H.Tahayori " Advances in Type - 2 Fuzzy Sets and Systems ",(2013), Springer Science +Business Media .
[19]. B.Hutton and I.Reilly ,"Separation Axiom in Fuzzy Topology Space" Fuzzy Sets and systems,Vol.3, (1980),pp..93-104.
[20]. Audi Sabri , Shihab Ahmed , "Analysis of the Reliability of Complex Systems by Using Fuzzy Falt Tree , Journal of Babylon University , Vol. (26) (2018) pp. 303-314.
[21]. A.Aygunoglu , V.Getkin , H.Aygun "An Introduction to Fuzzy Soft Topological Spacess" Hacettepe Journal of Mathematics and Statistics , Vol43(2), (2014)pp. 197-208 .
[22]. A.Kandil , O.A.E.Tantauy , S.A.EL.Sheikh , A.M.Abd EL-Latif , "Fuzzy Semi Open Soft Sets Related Properties in Fuzzy Soft Topological Spacess" Journal of Mathematics and computer Science .(JMCS) Vol.13 (2014).p,p(94-114).
[23]. A.Kandil , O.A.EL-Tantawi , S.A,EL-sheikh , Sawsan S.S.EL-Sayed " Fuzzy Soft Connected Sets in Fuzzy Soft Topological Spacess II" , Journal of Egypt Mathematics Society ,Vol.25(2017)pp.171-177.
[24]. A.Kandil , OAEL-Tantawy , S.A. AL-sheikh and A.M.Abd EL.Latif " Fuzzy Soft Semi Connected Properties in Fuzzy Soft Topological Apaces" Joural of the Eygption Mathematical Society , 25 ,(2017), p.171-177.
[25]. A.R.Roy , P.K.Maji "A Fuzzy Soft Set Theoretic Approach to Decision Making Problems" Journal of Computational And Applied Mathematics (Elsevler)(2007)pp.412- 418.
[26]. A.Sabrie and L.Abd AL.Hani "On Classification of Fuzzy Set Theory", Journal of Engineering and Applied Sciences,Vol.14,I,ssue14, (2019) pp..4786-4794.
[27]. A.Subrie , A.Musleh " Evaluating Fuzzy Reliability System Using Intuitionistic Fuzzy sets , " Journal of Babylon University Vol(6) No.(26), (2018) ,pp..313-321.
[28]. AL Swidi , L.A.Awad F.s.s "On Soft Turning Points" Baghdad Science Journal Vol 15(3) ,(2018),pp.(352-360).

[29]. C.Gunduz , S.Bayramov , "Some Results on Fuzzy Soft Topological Spacess" Mathematical Problems in Engineering , (2013), pp.(1-10).
[30]. K.T.Atanassov,"Intuitionistic Fuzzy Sets" ,Fuzzy Sets And Systems(20) (1986)pp.[87-96].
[31]. G.kalpana and C Kaloirani " Fuzzy Soft Topology " International Journal of Engineering Studies; Vol.(9)1 , (2017) , pp..45-56.
[32]. K.Borgohain , C.Gohain "Some new operations on Fuzzy Soft Sets" Journal of Modern Engineering Research (I J MER) , vol(4) (2014), pp. (65-68).
[33]. D.Molodtsov , "Soft Set Theory First Results" Computer and Mathematics with Applications , 37(1999) pp.19-31.

[34]. L.A.Al-swidi , S.A.Al Fathly "Types of Local Functions Via Soft Sets" International Journal of Mathematics Analysis , Vol .11(2017) no.6 , pp.255-265.
[35]. L.A.Zadeh , K.S.Fu , KTanaka and M.Shimura eds, "Fuzzy Sets and Their Applications to Cognitive and Decision processes" , (1975), Academic Press Inc.New York.
[36]. Lauy-A-AL-Swidi and Thu-Al-figar F.N. "a-density in Soft Topological Spacess" Journal of Babylon University, Vol(25), (2017), pp.1-13
[37]. Luay A.Al-Swidi , Mustafa H.Hadi "Charactizations of Continuity and Compactness with Respect to Weak Forms of w-open Sets" European Journal of Sientific Research Vol.57 No.4(2011) pp.577-882.
[38]. M.Borah , T.J.Neog , D.K.sut " A study on Some Operations of Fuzzy Soft Sets" . International Journal of Modern Engineering Research (IJMER) , Vol.2 (2012) , pp.(219-225).
[39]. P.k.Maji , R.Biswas and A.R.Roy , " Fuzzy Soft Set" J.Fuzzy Math. 9(3), (2001), pp. 587-602.
[40]. P.k.Maji , R.Biswas and A.R.Roy , "Soft Set Theory" Computers and Mathematics with Applications(45) (2003), pp.555-562.
[41]. S.Mishra and R.srivastava , "Fuzzy Soft compact Topological Spacess " Journal of Mathematics (2016), p.(1-7).

[42]. S.Roy , T.K.Samanta "A note on Fuzzy Soft Topological Spacess" , Annals of Fuzzy Mathematics and Information (a FμI) , Vol.3 ,No.2(2012) ,pp.305-311.

[43]. V.Cetkin , H.Aygun "A note on Fuzzy Soft Topological Spacess" 8th conference of the European Society for Fuzzy Logic and Technology (Eusflat), (2013) ,pp. (56-60).

[44]. Zorlutana , M.Akdag, W.K ,Min , S.Atmaca " Remarks on Soft Topological Spaces" Annals of Fuzzy Mathematics and informations Vol(3), No.(2). (2012), p.171-185 .

[45]. A.A.Ramedan "Smooth Topological Spaces " Fuzzy Sets and Systems (48),(1992),pp. 371-375.

[46]. AL Siwid LA , AL-Rabaye M.S "New Classes of Separation Axioms Via Special Case of Local Function" International Journal of Mathematical Analysis. vol.8(23)(2014)pp..1119-1131.

[47]. L.Zadeh, " Probability Measures of Fuzzy Events", Journal of Mathematical Analysis and Application, Vol.23, (1968),pp.421-427.

[48]. Audi Sabri and Luay Abd Al Hani "Finding and Taxonomy a new Fuzzy Soft points" , Journal of Engineering and Applied Sciences , Vol.14, issue.14, (2019),pp.4817-4821.

[49]. Audi Sabri and Luay Abd Al Hani "Soft Generalized Vague Sets: An Application in Medical Diagnosis" ,Indian Journal of Public Health Research & Development , Vol, No. 2 , (2019),pp.. 901-907.

[50]. B.M.Juk and Y.E.Jum " Fuzzy Topology and its Applications", Kangweon-Kynngki Math. Jour.(3),(1995),(1), pp..135-142.

[51]. B.P.Varal and H.Aygun "Fuzzy Soft Topology" Hacettepe Journal of Mathematics and statistics Vol 41(3) , (2012),pp..407-419.

[52]. B.Tanay , M.B.kandemir , "Topological Structure of Fuzzy Soft Sets", Computation And Mathematical with Application(61), (2011),pp..2952-2957.

[53]. D.N Georgiou , A.C.Megaritis and V.L Petro populous "On Soft Topological Spaces" App.Math. Inf.Sci, 7 No.5 , (2013) pp.(1889-1900).

[54]. I.Osmanoglu and D.Tolkat " On Intuitionistic Fuzzy Soft Topology" , Gen. Math. Notes Vol19 No 2, (2013) , pp..59-70.

[55]. J.Mahanta and P.K.Das "Results on Fuzzy Soft Topological Spaces" , (2012) ,arxiv:1203,0634.

[56]. Grzegorzewski, O.Hryniewicz "Soft Methods in Probability ,Statistics and Data Analysis" , (2002) ,Springer – Verling Heidelberg .

[57]. Luay A.A.H. Zahraa M.N, "e-Soft Separation Axioms in Soft Topological Spaces", Journal of Babylon University/Pure and Applied Sciences Vol(25)No.7(2017) pp..1-16.

[58]. Luay A.AL swidi and Fatima S.S Awad " On Soft Turning Point ", Baghdad Science Journal , Vol(15),(3),(2018)PP..352-360 .

[59]. Luay A.AL swidi and Maryam A.AL-Ethary , " Compactness with Gem-set" Int.Journal of math. Analysis, Vol.(8),No.23 ,(2014)pp.1105-1117.

[60]. M.Shabir and M.Naz. , " On Soft Topological Spacess" ,Computation and Mathematical with Application (6)(2011), pp..1786-1799.

[61]. S.E. Rodabaugh and E.P.Klement , "Topological And Algebraic Structures in Fuzzy sets" , (2003), Springer – Science + Business Media , B.V .

[62]. R.Erturk and M.Demirei "On the Compactness in Fuzzy Topological Spacess in Sostak's sense." Matematichki Vesnik 50, no. 3-4 (1998), pp. 75-81."

[63]. R.Lowen , "Compact Hausdorff Fuzzy Topological Spaces are Topological ",Topology and its applications" , Vol.12,(1981),pp..65-74.

[64]. S.Karatas , B.Kihe and M.Telliglu "On Fuzzy Soft connected Topological Spaces ,./.1 s" Journal of linear and Topological Algebra , Vol.(4) No.(3), (2015) pp.229-240.

[65]. Simsekler, S.Yuksel "Fuzzy Soft Topological Spaces" , Annals of Fuzzy Mathematics and Information Vol.5,No.1,(2013),pp..87-90 .

[66]. J.M. Mendel , "Uncertain Rule Based Fuzzy Systems . Introduction and New Directions" , (2017)Springer International Publishing AG.

[67]. B.X. Yao, J.L.Liu and R.X.Yan "Fuzzy Soft Set and Soft Fuzzy set " IEEE computer society , Fourth International conference on Natural computation . (4), (2008),pp.(252-255).

[68]. Gau W.L and Buehrer , D.J. "Vague Sets" .IEEE Transactions on Systems Man , and Cybernetics , 23, (1993), pp:610-614.

[69]. K.Alhazaymeh , N.Hassan " Generalized Vague Soft Set and its Applications" Vol. 77, No.3, (2012),pp.(391-401).

[70]. S.M.Chen "Fuzzy System Reliability Analysis Based on Vague Set Theory Proceeding of the (1997) IEEE International Conference on Systems". Man and Cybernetics , Orlando ,Florida ,USA, (1997), pp.1650-1655.

[71]. K.Alhazaymeh , S.A.Halim ,A.R.Salleh "Soft Intuitionistic Fuzzy Sets" Applied Mathematical Sciences , Vol.6,No.54, (2012),pp..2669-2680.

[72]. L.A.Zadeh , "Fuzzy Logic" ,Neural Networks and Soft communication of the ACM ,37(3),(1994)pp..77-84.

[73]. T.J.Noeg ,D.K.Sut , "On Union and Intersection of Fuzzy Soft Set" Journal Tech.Appl.Vol.(2),(5)(2011),pp..1160-1176.

[74]. B.Ahmed and A.Kharal , "On Fuzzy Soft Sets" , Advances in Fuzzy Systems ,(2009),pp..1-6.

[75]. N.Cagman , S.Enginoglu and F.citak, "Fuzzy Soft Set theory and its applications", Iranian Journal of Fuzzy systems, Vol.8, No.(3), (2011),pp..137-147.

[76]. Ahmad, B. and Kharal, A., " Fuzzy Sets, Fuzzy S-open and S-closed Mappings ", King Abdul Aziz University , Advances in Fuzzy Systems,2009, Article ID 303042, (2009)pp.(1-5).

[77]. Ali,S. M., "On Fuzzy Topological Vector Spaces", (2010), M.Sci., Thesis, AL-Qadisiyah University,.

[78]. Athar, M. and Ahmad, B., " Fuzzy Boundary and Fuzzy semiboundary", King Abdul Aziz University, Advances in Fuzzy Systems,2008, Article ID 586893, (2008), pp. 1-9.

[79]. Bourbaki, N., Elements of Mathematics, " General Topology", chapter 1-4, (1989),Spring-Verlog, Berlin, Heidelberg, New-York, London, paris, Tokyo, 2nd Edition.

[80]. Ali Ebrahimnejad and J. Luis Verlegay "Fuzzy Sets –Based Methods and Techniques for Modern Analysis "(2018), Springer International Publishing.

[81]. Dang, S., Behera, A. and Nanda, S., "On Fuzzy Weakly Semi-Continuous Functions", Fuzzy Sets and Systems,67(2), (1994),pp. 239-245.

[82]. Foster, D.H., " Fuzzy Topological Groups", J.Math. Anal. Appll, 67(2),(1979), pp. 549-564.

[83]. B.P.Varal,H.Aygun,"On Soft Hausdorff spaces" Annals of Fuzzy mathematics and Information's, vol.5, No.1,(2013), pp..15-24.

[84]. In, B.S., " On Fuzzy FC-Compactness", Comm.Korean Math.Soc. 13(1), (1998), pp. 137-150.

[85]. Kandil, A., Tantawy, O., Yakout, M. and Saleh, S.," Regularity and Separation Axioms on L-Fuzzy Topological Spacess ", J.Egyptian Math. Soc., 17(2), (2009), pp. 159-172.

[86]. F.Masuli, S.Mitra and G.Pasi , " Applications of Fuzzy Sets Theory", (2007),Springer –Verlag Berlin Heidlberg.

[87]. Mirmostafaee, A.K., "Fuzzy Topological Spacess ", In.S.S.S. of Iran, (2007), pp. 29-31.

[88]. Nouh, A.A., "On Convergence Theory in Fuzzy Topological Spaces and Its Applications", Czechoslovak Mathematical Journal, 55(2), (2005), pp. 295-316.

[89]. X.Liu, W.Pedrycz, " Axiomatic Fuzzy Set Theory and Its applications" , (2009),Springer – Verlag Berlin Heidlberg.

[90]. Pinchuck, A., " Extension Theorems on L-Topological Spacess and L-Fuzzy Vector Spaces" , (2001),M.Sci. Thesis, Rhodes University.,.

[91]. Rashid, M.H. and Ali,D.M., " Separation Axioms in Mixed Fuzzy Topological Spaces ", Rajshahi University, Journal of Bangladesh Academy of Sciences, 32(2), (2008), pp. 211-220.

[92]. Srivastava, R., Lal, S.N. and Srivastava, A. K., " Fuzzy Hausdorff Topological Spacess" , Banaras Hindu University, J.Math. Anal. Appl., 81, (1981), pp. 497-506.

[93]. Tang, X., " Spatial Object Modeling in Fuzzy Topological Spacess with Applications to Land Cover Change in China" , (2004) ph.D.Dissertation,University of Twente, Enscheda, The Netherlands.

[94]. R.H., Warren, " Neighborhoods, Bases and Continuity in Fuzzy Topological Spacess ", Nebraska University, The Rocky Mountain Journal of Mathematics, 8(3),(1978), pp. 459-470.

[95]. B. Liu, "Uncertainty Theory", Fifth Edition, (2018), Uncertainty Theory Laboratory,

[96]. S.A.Morris, " Topology without Tears " (version of October,(2017).

[97]. M.Matejdes, " Soft Topological Questions and Answers" international Journal of pure and applied Mathematics, vol.104, No.2, (2015) , pp..237-247.
[98]. E.E. Kerre, "Fuzzy topologizing with preassigned operations", International Congress for Mathematicians, Helsinki, (1978),pp. 61-62.
[99]. P.Mong. Pu, Y.M. Uu, "Fuzzy Topology I", J. Math. Anal. Appl. 76 (1980),pp. 571-599.
[100]. H. Ludescher, E. Roventa, "Sur les Topologies Floues Defines Al'Aidedes Voisinages", C. R. Acad. Sc. Paris t. 283 (1976) Serie A, 575-577.
[101]. R.H. Warren, "Fuzzy Topologies Characterized by Neighbourhood Systems" ,Rocky Mt. J. Math. 9 (1979),761-764.
[102]. M.E. Abd. El-Monsef, M.H. Ghanim, E.E. Kerre, A.S. Mashhour, "Fuzzy Topological Results", in: Proceedings Fifth Prague Topological Symposium(J.U. Novak, ed.) Helderman Verlag, Berlin, (1983), pp. 1-5.
[103]. M. Ghanim, E.E. Kerre, A. Mashhour, "Separation Axioms, Subspaces and Sums in Fuzzy Topology", J. Math. Anal. Appl. 102 (1984), pp. 189-202.
[104]. D. N. Georgiou and A. C. Megaritis, 'Soft Set Theory and Topology',Appiled General Topology, vol. 15, no. 1, (2014),pp. 93–109.
[105]. D. Wardowski, 'On a Soft mapping and its fxed points', Fixed Point Theory and Applications, (2013) ,pp. 1–11.
[106]. F. Lin, 'Soft Connected Spases and Soft paracompact spaces', International of Mathematical,computational,physical,Electrical and computer Engineering, vol. 7, no. 2,(2013) ,pp. 1–7.
[107]. G. Xuechong, 'A study on central Soft Sets : Definitions and basic operations', (2015).
[108]. M.I. Ali, F. Fengb, X.Liu, W. K. Min and M.shabir, 'On some new operations in Soft Set theory',Computer and Mathematics with Applications, vol. 57, (2009), pp. 1547–1553,.
[109]. M. I. Ali and M. Shabir, 'Algebraic structures of Soft Sets associated with new operations', Computer and Mathematics with Applications, Vol. 61, 2011 , pp. 2647-2654,.
[110]. N. Gagman and S. Enginoglu, 'Soft Set Theory and Uni-int

Decision Making', European Journal of Operational Research, vol. 207, (2010), pp. 848–855,.

[111]. N. Xie, 'Soft Points and the Structure of Soft Topological Spaces', Annals of Fuzzy Mathematics and Informatics, vol. x, no. x, (2015), pp. 1–14.

[112]. N. Cagman, S. Karatas and S. Enginoglu, 'Soft Topology', Computer and Mathematics with Applications, vol. 62, (2011), pp. 351–358.

[113]. S. Das and S. K. Samanta, 'Soft metric', Annals of Fuzzy Mathematics and Informatics, vol. 6, no. 1,(2013),pp. 77–94.

[114]. S. Hussain and B. Ahmad, 'Some Properties of Soft Topological Spaces', Computer and Mathematics with Applications, vol. 62, 2011 , pp. 4058–4067.

[115]. A.M.Ibrahim and A.O.Yusuf "Development of Soft Set Theory " , American International Journal of Contemporary Research , Vol(2), No.9 (2012) pp.. 205-210.

[116]. Fatma Adam and Nasruddin Hassan, " Q-Fuzzy Soft Set " , Applied Mathematical Sciences , Vol. 8, No.174(2014), pp..8689-8695.

[117]. Feng Feng, Changxing Li, Bijan Davvaz, M Irfan Ali ,"Soft Sets combined with Fuzzy sets and rough sets: a tentative approach" , Soft Computing(14)(2010), pp..899-911.

[118]. Abdul –Haddi .Z .Al_kha , " Using Some Mathematical Models in Reliability System" , 2006 , M.Sc. Thesis , Babylon University , .

[119]. Clifton A.EricsonII, "History of Fault Tree" ,(1999), The Boeing Company Seattle , Washington.

[120]. Douglas B. W. , "Graph Theory" , Second Edition ,(2007)Prentice-Hall of India Private Limited .New Delhi. .

[121]. S.Gao , Z.Zhang and C. Cao "Multiplication Operation on Fuzzy Numbers " , Journal Of Soft ware, Vol(4), (2009), No(4),.

[122]. Frankel , E. ,C. ,"System Reliability and Risk Analysis" ,2nd addition , 1987 ,printed in U.S.A. .

[123]. Guillaume Merle ,J. Roussed , J.Lesage ," Dynamic Fault Tree Analysis Based on The Structure Function ",(2011), IEEE , .

[124]. Jain S.P., Gopal K.,"An Improved of Selecting Network Topology for Optimal Terminal Reliability, 26 (1986) 2 .

[125]. Jianwen Xiang , Kazuo Yanoo . ,"Formal Static Fault Tree Analysis " ,(2011), IEEE ,.

[126]. Keiser E.G.,"Local Area Networks" , (1989),McGraw-Hill , Inc. , New York ,

[127]. Narsingh Deo ., " Graph Theory with Application to Engineering and Computers Since ", (2007), since Hallof IndiaPrivate Limited New Delhi , .

[128]. Nikolaos Limnios ,"Fault Tree ISTELTD ", (2007) .

[129]. OlexanderYerkin ," AN Improved Modular Approach for Dynamic Fault Tree Analysis ", (2011), IEEE.

[130]. Jerry M. Mendel, Robert I. BobJohn , " Type-2 Fuzzy Sets Made Simple " , IEEE Transactions on Fuzzy Systems , Vol(10), No. 2, (2002), pp. 117-127.

[131]. Srinath L.S. ,"ReliabilityEngineering ", 4^{th}edtion Affiliated East , (2005),Wast Press Private Limited -New –Delhi.

[132]. R.J.Wilson "Introduction to Graph Theory "(1996) ,Long Man Forth Edition.

[133]. Kai-Yuan. Cai " Introduction to Fuzzy Reliability ", (1996) ,Kluwer Academic Publisher.

[134]. W.E.Vesely , F.F. Goldberg ,N.H.Roberts and D.F.Hassl, "Fault Tree Handbook",V.S Nuclear ,Regulatory Commission Washing Ton , (2002), D.C.20555.

[135]. Neeraj Lata " Analysis of Fuzzy Fault Tree Using Intuitionstic Fuzzy Numbers ", (IJCSET), Vol(4)No.7 ,(2013) , pp..918-924.

[136]. G.S.Mahapatra , B.S.Mahapatra " Intuitionistic Fuzzy Fault Tree Analysis Using Intuitionistic Fuzzy Numbers ", International Mathematical forum , Vol(5), No21 , (2010), pp..1015-1024.

[137]. Salvatore Distefano and Antonio Puliafito "Dynamic Reliability Block Diagrams Vs Dynamic Fault Trees " ,(2007) , (IEEE).

[138]. Y.A.Mahmood.A.Ahmadi.A.K.Verma. A.Srividya .U.kumar " Fuzzy Fault Tree analysis : a review of concept and application " Int.J .syst .Assur. Eng.Manag , Vol(4) , No.1 ,(2013) , pp..(19-32).

[139]. Richard Banach , Marco Bolzzano " Retrenchment and the Generation of Fault Trees for Static , Dynamic and Cyclic Systems "

Work Partly Supported by the E.U.ISAAC Project ,Contract no.AST2-CT-2003, pp..1-14 .
[140]. Liping He, Zongwen An: " Possibilistic Information Measure of Importance in Fault Tree Analysis " , International Conference on Fuzzy Systems and Knowledge Discovery (FSKD 2010), pp.. 487-491.
[141]. Charles E. Ebeling " An Introduction to Reliability and Maintainability Engineering " , (2007),Tata McGrraw- HillPublishing Company Limited , 7nth edition.
[142]. Audi Sabri and Ali Salman , " Sequential Procedure to Derive Minimal Cut Sets for a Complex System Directly From Its Fault Tree " , Babylon university Magazine 2 (2014) 22 .

[143]. P.Anglov, S.Sotirov " Imprecision and Uncertainty in Information Representation and Processing " , (2016), Springer International Publishing.
[144]. Marko Cepin, Borut Mavko , " A dynamic Fault Tree " , Reliability Engineering and System Safety (75),(2002) pp..(83-91)

[145] .Sharma Anurag , Jha Manoj ,Qureshi M.F , " Reliability Investigation of Series – Parallel and Components of Power System Using Interval Type -2 Fuzzy Set Theory " , International Journal of Innovative Research in Science ,Engineering and Technology , Vol(3) ,No.11 ,(2014) , pp.17676-17687 .

[146]. Z.Ma , F.Zhang , L.Y.j.Cheng "Fuzzy Knowlledge Management for the Semantic Web" (2014),Springer Science + Business Media ,.
[147]. Tridiv J yoti Neog, Dusmanta Kumar , " Theory of Fuzzy Soft Sets from a New perspective " ,Int.J Latest Trends Computing Vol.(2), No(3), (2011), pp.439-450.
[148]. Çigdem Gunduz(Aras), Hande Poşul , " On Some New Operations In Probabilistic Soft Set theory " , European Journal of Pure and Applied Mathematics , Vol.(9), No(3), (2016), pp.333-339.
[149]. Neeraj kumar Goyal , Ravindra Babu Misra and sanjay kumar chaturvedi ,"SNEM: A new Approach to Evaluate Terminal Pair Reliability of Communication Networks " , Journal of Quality in Maintenance Engineering Vol(11), No.3 ,(2005),pp..239-253.

[150]. James J. Bukckley :Fuzzy Probability and Statistics" (2006),Springer Verlag Hiedeberg.
[151]. Shshank chaube , S.B.Singh , " Fuzzy Reliability Theory Based on Membership Function" International Journal of Mathematical , Engineering and Management Sciences, Vol(1) , No.1,(2016), pp..34-40.
[152]. H.A.Ali , M.Radi "Fuzzy Conditional Reliability for Some Continuous Distributions "Journal of Babylon university Pure and Applied sciences , Vol(21) , No.6 , (2013), pp..2001-2007.
[153]. W.Kosinski , PP.rokopowicz and D.stezak "Calculus with Fuzzy Numbers" Springer-verlag Berlin (2005), pp..21-28.
[154]. Wang peizhuang "Theory of Fuzzy Sets and Its Applications" Shanghai Science and Technology Press,(1983),(chinese).
[155]. Ahmed Shawky Moussa " Soft Probability: Motivation and Foundation" 2003,International IPSI-(2003) Conference In Montenegro.
[156]. Marko Cepin "Assessment of Power System Reliability " , (2011), Sipringer-Verlag London limited .
[157]. Amit Kumar, Shiv Prasad Yadav, and Surendra Kumar "Fuzzy Reliability of a Marine Power Plant Using Interval Valued Vague Sets" , (2006) Chaoyang University of technology, ISSN 1727-2394.
[158]. D . Pandey , S.K.Tyagi , and Vinesh Kumar " Reliability Analysis of a Series and Parallel Network using Triangular Intuitionistic Fuzzy Sets " An International Journal (AAM) Vol.6, (2011) pp 105-115.
[159]. Al-Ali A.A and Subri U.A " Evaluation of Reliability and Main Time to Failure of Power Plant" Journal of Babylon university, Vol.7 No.3,(2002),Iraq
[160]. Blacket D.N. , "Elementary Topology" , 1983 , Acad –emicpress ,Inc. ,London .
[161]. ,"Fault Tree and Book with Aerospace Applications" , (2002) ,NASA office of Safety and Mission Assurance .
[162]. Lars H. , " Safety Analysis Principles and Practice in Occupational Safety" , (2011) , British library of Congress , Cataloging in Publication Data .
[163]. Sandler G. ,"System Reliability Engineering", (1963), prentice – hall Englewood Cliffs.
[164]. A.S.Moussa "Questions on Probability Theory and its Effect on Probabilistic Reasoning " , (2003) ,7th Joint Conference on Information Science , Cary, NC, USA.

[165]. D.Dubois and H.Prade, " Fuzzy Sets and Probability: Misunderstandings, Bridges, and Gaps", (1993) ,in proc. Of the 2nd IEEE International Conference on Fuzzy Systems. San Francisco, CA, USA.
[166]. Tohn Yen and Reza Langari ,"Fuzzy Logic Intelligence , Control and Information" , (2007) , Published by Dorling Kindersley (India) Pvt Ltd.
[167]. Wang Limin," Fault Tree Analysis For Oil Tank Fire and Explosion", 2010 , IEEE.
[168]. S.M.Chen, "Analysing Fuzzy System Reliability Using Vague Set Theory" International Journal of Applied Science and Engineering,(1), (2003), pp(82-88).
[169]. E.Nikolaidis,D.M.Ghioce and S.Singhal, "Engineering Design Reliability Hand book", (2005), Copyright by CRC press .

Printed by Books on Demand GmbH, Norderstedt / Germany